Materials Forming, Machining and Tribology

Series editor

J. Paulo Davim, Aveiro, Portugal

More information about this series at http://www.springer.com/series/11181

J. Paulo Davim
Editor

Ecotribology

Research Developments

Springer

Editor
J. Paulo Davim
Department of Mechanical Engineering
University of Aveiro
Aveiro
Portugal

ISSN 2195-0911 ISSN 2195-092X (electronic)
Materials Forming, Machining and Tribology
ISBN 978-3-319-36191-8 ISBN 978-3-319-24007-7 (eBook)
DOI 10.1007/978-3-319-24007-7

Springer Cham Heidelberg New York Dordrecht London

Softcover re-print of the Hardcover 1st edition 2016

Printed on acid-free paper

Springer International Publishing AG Switzerland is part of Springer Science+Business Media
(www.springer.com)

Preface

In the recent years, ecotribology (or environmentally friendly tribology) has gained increasing importance in green science and engineering. It is current report ecotribology or green tribology, as environmentally acceptable tribological practices, namely *savings of resources of energy and materials, optimizing product usage and design, reducing energy consumption and the impact on the environment.* Today, it is normal to include several topics under the umbrella of ecotribology, namely biomimetics surfaces, control of friction and wear, biolubricants, environmental aspects of lubrication and surface modification techniques as well as tribological aspects of green applications, such as wind-power turbines, or solar panels.

The purpose of this book is to present a collection of examples illustrating review studies and research in ecotribology. Chapter 1 of the book provides ecotribology development, prospects, and challenges. Chapter 2 is dedicated to advancements in ecofriendly lubricants for tribological applications (past, present, and future). Chapter 3 describes new emerging self-lubricating metal matrix composites for tribological applications. Chapter 4 contains information about multi-objective optimization of engine parameters with biolubricant-biofuel combination of VCR engine using Taguchi–Grey approach. Chapter 5 describes biolubricants and potential of waste cooking oil. Finally, Chap. 6 is dedicated to two-body abrasion of bamboo fiber/epoxy composites.

The present book can be used as a research book for a final undergraduate engineering course or as a topic on tribology at the postgraduate level. Also, this book can serve as a useful reference for academics, researchers, mechanical, materials and industrial engineers, professionals in tribology and related industries. The scientific interest in this book is evident for many important centers of research, laboratories, and universities as well as industry. Therefore, it is hoped this book will inspire and enthuse others to undertake research in ecotribology.

The Editor acknowledges Springer for this opportunity and for their enthusiastic and professional support. Finally, I would like to thank all the chapter authors for their availability for this work.

Aveiro, Portugal J. Paulo Davim
October 2015

Contents

Chapter 1
Ecotribology: Development, Prospects, and Challenges

Ille C. Gebeshuber

Abstract Ecotribology is gaining increasing attention. Our view of the environment has changed from regarding it as a constant that provides resources and acts as a sink for waste toward a more complex view, where the environment is seen as a variable that can be influenced by our activities and on which we are utterly dependent. Ecotribology can be seen as the answer to this changed role of the environment. In the very word *ecotribology* economical and ecological aspects meet, and indeed the field comprises green tribology, sustainability, ecological aspects, economical aspects, environmentally compatible lubricants, environmentally friendly tribology, tribology of eco-friendly applications, tribology for energy conservation, tribology for life, and renewable energy tribology. This chapter deals with components, goals, optimization levers, challenges, and prospects of ecotribological systems and gives ample examples in which regard we can learn from living nature via biomimetic approaches to achieve efficient ecotribology, concerning materials, structures, and processes.

1 Introduction

Ecotribology is gaining increasing attention. Our view of the environment has changed from regarding it as a constant that provides resources and acts as a sink for waste toward a more complex view, where the environment is seen as a variable that can be influenced by our activities (cf. industrialization, species extinction [8, 9], global challenges [43]) and on which we are utterly dependent. Peter F. Jost

I.C. Gebeshuber (✉)
Institute of Applied Physics, Vienna University of Technology, Wiedner Hauptstrasse 8-10/134, 1040 Vienna, Vienna, Austria
e-mail: gebeshuber@iap.tuwien.ac.at

I.C. Gebeshuber
Aramis Technologies Sdn. Bhd., 14 Jln. BK 5D/1C, 47180 Bandar Kinrara, Puchong, Selangor, Malaysia

J.P. Davim (ed.), *Ecotribology*, Materials Forming, Machining and Tribology,
DOI 10.1007/978-3-319-24007-7_1

said at the 4th World Tribology Congress in Kyoto, Japan, in 2009 that a focus on tribology might give “breathing space” while comprehensive solutions to environmental problems were being addressed and suggested that tribology must fall in line with the major politics of world environment and energy [6]. Ecotribology can be seen as the answer to this changed role of the environment. In the very word *ecotribology* economical and ecological aspects meet, and indeed the field comprises green tribology, sustainability, ecological aspects, economical aspects, environmentally compatible lubricants, environmentally friendly tribology, tribology of eco-friendly applications, tribology for energy conservation, tribology for life, and renewable energy tribology.

In 2006, Bartz published a paper in Tribology International where he gave in dense bullet form environmentally acceptable tribological practices (see Sects. 3.2 and 3.3 for related further elaboration). Sasaki in 2010 published a paper called “Environmentally friendly tribology (Ecotribology).” Sasaki sees the need for ecotribology based on global warming and increased pollution and suggests multiscale texturing and diamond-like carbon (DLC) coatings for green lubrication as prospects for ecotribology [78]. In the 2012 book “Green Tribology: Biomimetics, Energy Conservation and Sustainability” [66] specific focus is taken on biomimetic surfaces, materials, and methods as well as green and sustainable lubricants and materials, as well as applications.

This chapter takes the whole approach further, and introduces in Sect. 4 the concept of biomimetic metal management for tribology, inspired by materials, structures, and processes in living nature.

2 Motivation

Humans are one of the most successful, if not the most successful species on Earth. Our numbers rise and rise, and the rate of technological advancement is astonishing. If one looks at the age of the Earth compressed into a time span of 1 year, with today being midnight, December 31, the industrial revolution has only started 2 s before midnight. During this relatively short time, we have developed internal combustion engines, computers, spaceships, and chess- and table tennis playing robots. As many benefits as this development has, there are various drawbacks that increasingly cannot be ignored anymore. Pollution, habitat destruction followed by species extinction on a massive scale, climate change, and increasing problems with resistant microorganisms directly result from our current way of resource management and doing engineering. The global challenges [42] cannot be ignored anymore, as cannot be ignored that we might have brought our planet close to a tipping point [8] and have unleashed the sixth mass extinction of species [9], with mass extinction being defined that 80 % of all species go extinct. The last mass extinction of species was 55 millions of years ago, when the dinosaurs disappeared.

In about 1760, when coal started to be used as an energy source (before only renewable natural energy was used), only 2 s before midnight (which represents today) in the condensed view introduced above (Fig. 1), the industrial revolution started—and with it the current rise of Carbon dioxide in the atmosphere. That time, production methods started to transit from handmade to machine-made. Another major change took place in the mid 1850s, with the beginning of commercial drilling and production of petroleum.

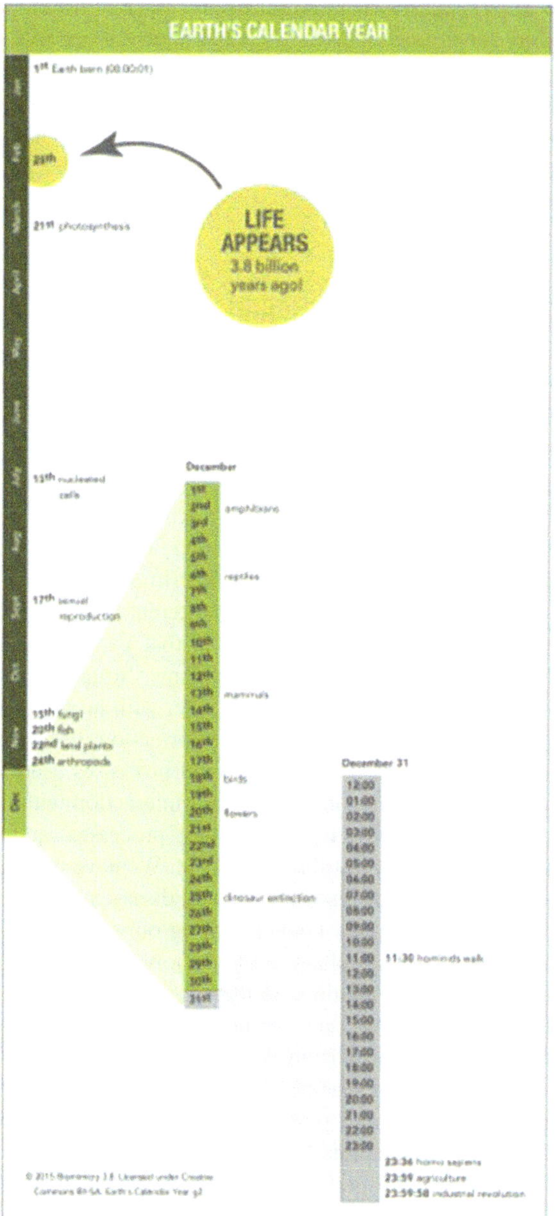

Fig. 1 History of the Earth condensed to 1 year. © 2015 Biomimicry 3.8 licensed under creative commons BY-SA

Increasingly replacing current technologies and approaches with sustainable and green technologies might substantially contribute to successfully address these problems and challenges. We need to take action. In our current world, where economic aspects are of paramount importance, ecological challenges can only be addressed when the approach is economically fruitful. And exactly here the concept of ecotribology, in which ecological and economical aspects are interwoven, becomes so important. Another advantage of tribology as opposed to various other, more theoretical approaches, is its inherent application orientation. Prof. Jost, the president of the International Tribology Council, mentioned in his opening address at the World Tribology Congress 2009 in Kyoto that a focus on tribology might give breathing space while fuller solutions to environmental problems are being addressed. Tribology must fall into line with the major politics of world environment and energy.

3 Efficient Ecotribological Systems

3.1 Components of Efficient Ecotribological Systems

Efficient ecotribologial components (Table 1) are ecologically and economically justifiable. The introduction of a concept of the *cost* of our current approaches for future generations would cause a shift in our perception, and might leapfrog our society toward sustainability. If one watches the web of life through time as an optimized tribosystem (thought experiment) the first interesting aspect that comes to mind is the very low amount of different materials that are used. *Structures rather than material* is a deep principle that can be extracted when analyzing plants, animals, humans, and microorganisms. Structures from the living world (such as bone and wood) are highly elaborate, with up to 11 hierarchical layers, with added functionality on each layer. The information for the materials, and the structures, as well as for the surfaces, agents, and processes used in living nature is inscribed in genomes, in combination with their interaction with the environment. Development of a similar approach to machines, processes, and systems in our technological world might be worth thinking about. We have just started to develop machines that self-assemble, self-repair, warn, and disassemble when needed or wanted. We have (apart from computer codes) nothing only slightly similar to a genetic code in our machines. No resemblance of phenotype and genotype, let alone developmental plasticity and adaptations to the environment, and also to their changes with time. Programmable materials (without emergent dangerous properties) might be a very interesting and promising way to start to address our current problems.

Slowly we are starting to develop synthetic materials that sense and respond to external stimuli and react to environmental parameters such as temperature and chemical composition, by changing their chemistry and/or shape, and thereby their functionality and performance (see Soft Matter Special Issue “Reconfigurable soft

Table 1 Components of ecotribological systems

Category	Importance	Challenges for R&D, in brackets selected exemplary publications treated in detail in the text and in Table 2
Materials	Medium	Optimization of shape rather than material
		Use of a limited amount of materials, slightly chemically varied and smartly structured Less use of metals [56]
		Tailored materials with response to the conditions of the tribosystem, as good as necessary, not as good as possible
		Benign materials
		Biomined metals (see Sect. 4.1)
		Recyclable materials
		Materials with an *expiration date*
		Triggered materials [82]
Structures	High	Smart structures
		Reactive structures [82]
		Biomimetic functional structures (see Sect. 4.2)
		Hierarchical structures
		Time-limited arrangements [95]
		Programmed structures [85]
		Self-assembled structures
		Directed assembly of structures [15]
		3D printed structures
		Foldable structures [1]
Surfaces	Medium	Structured surfaces [1]
		Hierarchical surfaces
		Texture rather than material
		Coatings
		Additive layers
Agents	High	Physical and chemical properties
		Effect on the environment
		Effect on organisms [64]
		Temporal changes of properties [40]
		Changes of properties in the triboprocess [89]
Processes	Medium to low	Energy efficiency
		Chemically guided synthesis of functional structures and materials (see Sect. 4.3)
		Fraction of process relevant energy, destructive energy and waste as well as recyclable energy
		Efficiency of reusability of process energy

matter", edited by Balazs and Aizenberg [7]). The coupling between mechanical, electrical, optical, or thermal behavior in such reconfigurable materials promises exciting applications in ecotribology.

In the following, we introduce eight key papers from the above-mentioned special issue, relate them to components of ecotribological systems (Tables 2 and 3) and highlight potential advantages from such new approaches. The methodology of this approach is inspired by the transdisciplinary knowledge integration method [45, 51] and innovision, a method developed by the author of this chapter [29, 30]. "Innovision is a new method to do research and to think. The path from innovation to innovision is characterized by the development of a new framework of thinking that is the prerequisite for the provision of solutions."

The author has long-standing experience in tribology, especially in micro- and nanotribology. Her work is on the global challenges for humankind and how humankind can address them with sustainable, disruptive technologies, accessing and providing human knowledge and educating future generations. She received her formal science and engineering education in Europe, spent some postdoctoral research time at the physics department of the University of California in Santa Barbara and has been living and working in Southeast Asia since 2008. With a team that consists of architects, designers, engineers, biologists, artists, economists, and a veterinary scientist they learn from virgin forest ecosystems and from each other, jointly developing and refining new synergistic approaches, blending Western and Eastern ways of thinking and perceiving the world and current challenges [30].

The results of the knowledge integration are presented below and in Tables 2 and 3.

Table 2 Components of ecotribological systems: Inspiring publications

Paper title	Category	Keywords and concepts of interest for ecotribology
Autonomic composite hydrogels by reactive printing: materials and oscillatory response [56]	Materials	Less use of metals
Design of polarization-dependent, flexural-torsional deformation in photo-responsive liquid crystalline polymer networks [82]	Materials	Triggered materials
	Structures	Reactive structures
Photoactivatable CRISPR-Cas9 for optogenetic genome editing	Materials	Change of tribological properties with external signals
Tunable shape transformation of freezing liquid water marbles [94]	Structures	Time-limited arrangements
Bioinspired materials that self-shape through programmed microstructures [85]	Structures	Programmed structures
Reconfigurable assemblies of Janus rods in AC electric fields [15]	Structures	Directed assembly of structures
Shape-responsive liquid crystal elastomer bilayers [1]	Structures	Foldable structures
	Surfaces	Structured surfaces
Reconfigurable and actuating structures from soft materials [40]	Agents	Temporal changes of properties
Fluid-driven motion of passive cilia enables the layer to expel sticky particles [89]	Agents	Changes of properties in the triboprocess

Table 3 Aspects of efficient ecotribological systems, their importance, and examples of how these goals are achieved in living nature

Main aspects		Negative impact on environment	Examples from living nature
Raw materials		High (mining)	Few different chemical elements are used in organisms. The human body for example consists of mainly six chemical elements: 65 % of its mass is Oxygen, 18 % Carbon, 10 % Hydrogen, 3 % Nitrogen, 1.5 % Calcium, and 1.2 % Phosphorus. Of all other chemical elements human bodies contain 0.2 % or less.
	Harmful emissions	High (pollution)	Few
	Water consumption	High (global challenge 2) [42]	Drinking, transport of seeds, living environment (for fish and others)
	Waste water	High (planetary boundaries) [74, 84]	Full of food and fertilizer for others
	Exploitation of resources		In most cases closed circles
	Accessory material		Rarely
	Waste		Rarely
Material transport	Harmful emissions	High	In soil, in air, by water, on animals
	Energy consumption		In living nature, rarely material transport takes place for the production of products. Most is locally harvested and free.
Production using materials			Few base materials with slight chemical variations and amazing structure-function relationships, often structure rather than material causing the respective functions
	Accessory materials		Very rarely, and if used, they are recycled and reused internally
	Harmful emissions		Oxygen was once the toxic gas produced by plants—the web of life adapted and now this gas is important for survival in most organisms—opportunistic approaches in living nature are omnipresent
	Energy consumption		Low
	Solid product waste		Urine and faeces are used as food or fertilizer by other organisms, as are plant shells and nutcases, and other equivalent materials

(continued)

Table 3 (continued)

Main aspects		Negative impact on environment	Examples from living nature
	Special waste		Food or fertilizer
	Recycling		Omnipresent recycling on all levels
	Liquid product waste		Fertilizer
	Water consumption		Low
	Wastewater		Urine—fertilizer
Product transport	Energy consumption	High	By air and water, free
			Seed transport
			Migrations
			Homing of fish such as salmon and turtles when reproducing
	Harmful emissions		Little, some natural gases (sometimes toxic) emitted from plants and animals
Usage	Harmful emissions		Sometimes—like when the first archaic plants released oxygen—killing nearly all the other organisms on Earth—with time, new organisms arose that actually vitally need oxygen
	Energy consumption		Low, but not the factor that organisms are optimized for [20]
	Packing waste		Biodegradable (fruit peel, nut shells)
Disposal	Combustion		Some organisms need burnt ground to thrive (e.g., fire beetles, [77])
	Recycling		Omnipresent, dead organisms, and waste of organisms are used as food or fertilizer
	Re-usage		Diatom frustules, hermit crab shell
	Utilization		Remnants of certain animals are used as building materials in certain animals (e.g., hermit crabs) or as shelter
	Dumping ground		Limestone hills (remnants from corals), coal deposits, white cliffs of Dover (coccolithophorid shells), etc.

Main aspects from [10]

In the category "Materials", printed autonomic hydrogels with swell/deswell behavior depending on external conditions could potentially be used as valves, actuators, delivery units, or micromachine material [56] and thereby result in less use of metals in machines and devices. Patterned films that twist and fold in response to temperature changes, made from shape-responsive liquid crystal elastomer bilayers [1] could, for example, manage lubricant supply and leakage, and provide valuable progress to the categories "Structures" and "Surfaces". Changes in lubricant quality with time can be buffered by reconfigurable machine material, switching the demands from rather illusory lubricants that never age to the development of machine parts that react to the quality of the lubricant. In this way, the same tribosystem could also be used in various environments (such as moderate zone, the tropics, the arctic) with specific lubricants tailored for the local needs, without having to change machine components or materials (inspiration from [40]). Fluid-driven motion of passive hairs that enable expelling of sticky particles [89] could potentially help in removal of wear particles and prevent unwanted layer formation during the triboprocess (category "Agents" in Table 3). The work by Smith and coworkers on polarization controlled flexural–torsional deformations in materials with complex geometry, boundary conditions, and loading conditions can give valuable input regarding the management of tribological issues in morphing structures (e.g., structures that can turn on or off enhanced mechanical functions like elastic instabilities from bistable arches) or in soft robotics [82]. Zhang and coworkers [94] experimented with coating small amounts of water with nanoparticles of varying hydrophobicity and the resulting shape transitions during freezing and remelting cycles. Inspired by this approach, the development of fast and easy production ways for tailor-shaped particles for tribological applications can be envisaged (e.g., for standardized test or in managing freezing on windshields of cars and airplanes). Rikken and coworkers [73] review the manipulation of micro- and nanostructure motion with magnetic fields. This opens exciting opportunities to align nanoparticles and/or to induce shape changes that make them anisotropic, resulting in interesting applications for tailored reversible tribological properties (e.g., higher or lower friction coefficient depending on specific demands) and also new in situ tracking methods based on confocal microscopy and dynamic light scattering that allow for real-time measurement of the motion of nanostructures. Anisotropy is widely represented in biological materials, but still not well developed in current technological devices. Especially hard–soft transitions pose problems. The work by Chaudhary and coworkers [15], inspired by Glotzer and Solomon [44], deals with microscale particles that assemble reconfiguably in AC electric fields (tunable patch type, size, and location) and that could become the "atoms" and "molecules" of tomorrow's materials.

Just recently, Nihongaki and coworkers reported in the scientific journal Nature Biotechnology about their development of gene switches that are activated by light. In this way, precise external signals can control which genes are active and which not. The group aims to expand the work to different colors activating different genes [64]. Thinking such approaches further one might envision programmable materials that can be controlled by external signals. One potential application in tribology might be tunable surface roughness of tires that correlates with the roughness of the ground the vehicle is navigating on, resulting in different tire properties on wet streets at the onset of rain (when due to dissolved fuel components the risk of hurling is largest) and on stony roads where different tire properties would be favorable to ensure the safety of the device.

3.2 Goals of Efficient Ecotribology

The goals of effective ecotribological systems are in the production of the raw materials, in their transport and subsequent product production, in their usage and finally, in their disposal (Fig. 2, [10]). These goals are located in three main areas: production (agents, machine components, systems), reaction (agents, products, direct effects of waste agents) and the life cycle of products (impact on the environment during application and during degradation). Minimum contamination and impact on the environment combined with economical feasibility shall be one of the major goals and aspects of ecotribology (Table 3).

Many of the major challenges that humankind is currently facing are based on our current unsustainable ways of doing engineering, business, and marketing. Pollution, the piling waste problem and the new major water shortages we are

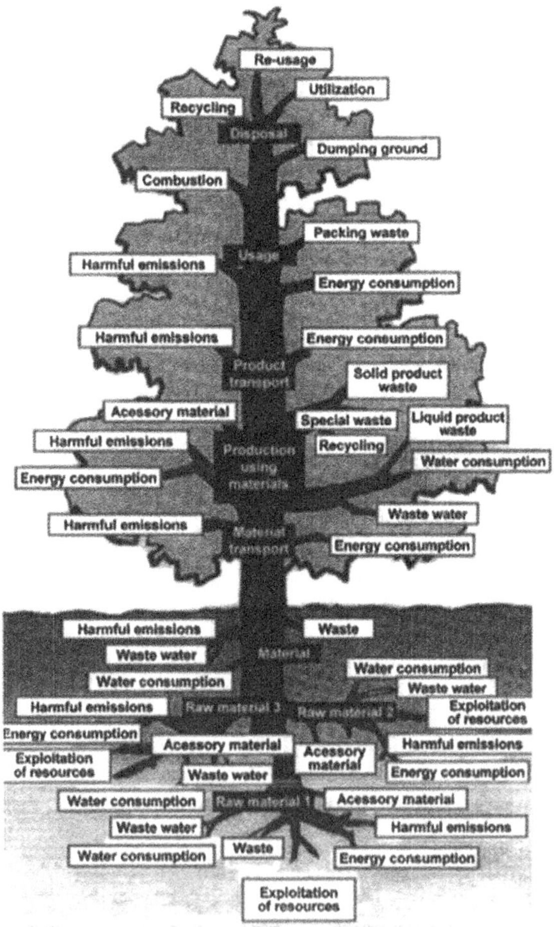

Fig. 2 Eco-Balance tree. Image © Elsevier [10]. Image reused with permission

facing are all relatively new problems: many of them were initiated by the start of the industrial revolution, at the end of the eighteenth century, when new manufacturing processes enabled the transition from handmade to machine-made products. At this turning point in history, major economic and social changes took place, and production and use of machines, mining methods, and modes of transportation drastically changed. With the industrial revolution, metals, fossil fuels, and concrete started to gain in importance, leading to the engineering-related problems we are facing now.

During the last few hundreds of years, it increasingly turned out that the continuously increasing usage of these materials is unsustainable and needs to be addressed rather sooner than later with novel, potentially disruptive ways of doing engineering, providing similar quality of life, but not at the expense of the biosphere.

New ways of teaching, education, resource management, engineering, economy, etc. might provide the intellectual background and proactive groups of people to deeply think about current problems, develop potential ways to address them, and work on the realizations of solutions.

However, we should not try to just superficially implement strategies such as sustainability and recycling in our new approach. This does not work. A new, disruptive, sustainable approach to engineering, and, in tribology, sound ecotribology need to result from the new complex system we create, as an emergent property; such as collaboration emerges in economies, although all the actors are just concentrating on their own benefit. This kind of emerging sustainability, emerging resource cycling might be a viable solution to current problems.

There is one proven sustainable system that deals with resources, transport, production, use, and disposal of huge masses: living nature. By learning from living nature we might develop ways to improve the current status of the biosphere. Nature can teach us deep principles and viable approaches, and also how not to do things. Additionally to living nature, people and their engineering have the marvelous power to choose their materials. Organisms cannot easily change the materials they use, e.g., for building up their skeletons. Some are stuck with exotic materials such as the mineral celestite (Strontium sulfate), as is the case for the marine organisms with the name Acantharia (Sect. 4.3.1, [28]). They cannot simply switch to another material. People, however, in their technological attempts, can.

The oil dependency of our industrialized nations is still very high: T.C.M. Gupta from Lube World Holdings in Kuala Lumpur Malaysia mentioned in his keynote address at the Malaysian International Tribology Conference 2013 the oil consumption by end use in the energy sector: electric power 3 % (3 % oil dependent), residential & commercial, 6 % (21 % oil dependent), industrial 24 % (43 % oil dependent), and transportation 67 % (66 % oil dependent). In organisms, on the other hand, material transport and product transport are performed without reliance on fossil fuels (apart from the interesting species of oil consuming microbes, see [13]). Organisms only use local resources.

Goals of efficient ecotribology can be achieved by learning from nature. Bioinspired transport in ecotribology is addressed below, for further examples, see

Sect. 4. If we could start to develop engineering approaches that would help us departing from the transport intensive way we are currently using, major energy savings would result, and benefits for the whole biosphere. One potential approach is to emulate the local resource management approach of living nature, and to increase the use of 3D printers using only a couple of materials (similar to living nature), slightly chemically varied (this could potentially be done by the end user?!), resulting in structurally elaborate functional prints (for various applications). The base materials for the 3D printers of course need to be produced and delivered, but since most people in settlements would use similar or the same materials, whole new ways of transport, storage, maintenance, etc. could be envisaged—similar to current water and electricity supplies. Perhaps even local resources can be the major materials of choice, strengthening local markets, materials, and people. 3D printing is a booming technology, allowing for rapid prototyping of complex structures and also for printing of chemical molecules and tissues with a commercially available 3D printing platform: We have already started to initiate chemical reactions by printing reagents directly into a 3D reactionware matrix, and the future might bring printing capabilities for whole organs or machines, from basic ingredients, simple base materials, in our homes [54, 86].

3.3 Optimization Levers in Ecotribology

Inspired by the four optimization levers in tribology suggested by Scherge and Dienwiebel [80] and by the eco-balance tree suggested by Bartz [10] (Fig. 2) we propose focusing on the following in the development of a sustainable ecotribology concept: material selection, finishing, additives, and running-in need to be optimized with regard to the used materials, structures, and processes. Material transport, production using materials, product transport, usage, and finally disposal need to be tailored to prevent further exploitation of resources and negative impact on the environment. Biomimetics can also here give valuable inspiration. Plants, animals, and microorganisms utilize metals in completely different ways than humans (see Sect. 4). Exquisite tribological properties are achieved via water-based chemistry in a state of dynamic nonequilibrium, subject to limits and boundary conditions. As tribological contacts are required to operate under greater tolerances and more extreme conditions than organisms, only principle transfer is possible and advisable in biomimetic tribology, not copying.

"Conventional technological lubricants are uniform chemical compounds achieving specific results regarding the physics of the tribosystem. Biological lubricants are mainly water-based, and in many cases the lubricant chemically attaches to the surface (such as in the lubricant layers reported for synovial joints, the lung, or the eye). Current manmade lubricants are mainly oil based. One reason for this is the thermal instability of water-based lubricants at elevated temperatures. One promising area for bioinspired water-based lubricants are ceramic MEMS that work at ambient conditions, with the lubricants chemically attaching to the surface, building monomolecular lubricant layers [90]" [26, 27].

3.4 Challenges for Efficient Ecotribology

More often than not, ecological and economical requirements do not overlap too much. It is the art of good ecotribology to establish a balance that allows this field to thrive. Especially, when collaborating with huge companies (international corporations such as big oil firms) the visions and mission gaps between the various stakeholders might seem unbridgeable. Corporations have goals and underlie certain constraints that arise from their complex structure. These goals differ from goals of society, and from goals of research. Value-based science is a complicated field, since besides research and development ethics, society and politics are important stakeholders [2, 3]. As Gebeshuber [30] pointed out in "Value based science: what we can learn from micro- and nanotribology", tribology can provide first interesting solutions in this regard. Tribology has always been about properly managing fuel consumption, wear and tear, and optimizing machines, devices, and lubricants. One might even go as far (as stated by Prof. Franek, Scientific Director of the Austrian Center of Competence for Tribology) and say that tribology is inherently green.

The development in tribology regarding policies and research foci is promising: being aware of the damage that has been done with engineering that mainly ignored the environment and just focused on technological improvements (faster, cheaper, smaller, etc.), current funding bodies and tribology research institutions, groups and single researchers and thinkers increasingly incorporate ecotribology thought in their approaches.

Ecological requirements on technological developments need to be put in a concise perspective. We are too much rooted in the present with our regulations and policies. Here, some inspiration could come from indigenous peoples who in their actions think about the seventh generation after them: The Great Law of the Six Nations of the Iroquois, a northeast Native American confederacy, requires that every deliberation considers the impact of any of its decisions on the next seven generations [52]. Indeed, one of the global challenges identified by the Millennium Project is exactly in this realm: Global Challenge 5 deals with the relationship of policymaking and the sensitivity to long-term perspectives: "How can decision-making be enhanced by integrating improved global foresight during unprecedented accelerating change?" [43]. Short work contracts, short research projects, increased frequency of switching jobs, low identification with major goals, and low feeling of being responsible contributes to challenge 5, and need to be properly addressed to go beyond selfish, short-term economic interests. In this respect, the reader might be interested in Berne's conversations with nanoscientists. Berne argues that scientists who are trying to catch "the wave" of nanotechnology have little motive or opportunity, let alone incentive, for reflection [11, 47].

Short-term thinking increasingly is important and demanded in science and technology. Numbers need to be met, and few researchers have time, capabilities, and support to draft long-term ideas and concepts. Indeed, how policymaking can be made more sensitive to global long-term perspectives is one of the global challenges identified by the Millennium Project [43]. Living nature seems to have implemented various approaches to address long-term perspectives, written in the DNA of living beings [81].

3.5 Prospects of Efficient Ecotribology

Given the huge amount of papers that are published every year, with many of them neither being read nor cited [48], we need a system that allows everyone, including key stakeholders such as industry, fellow scientists (from the same and other fields), and the general public access to research results [34]. Open Access publications are a start, but expensive. Policies and standards and regulations are important in tribology, especially for product and agent development. The policy makers need to stay informed about potentially hazardous outcomes of research and development, especially in multidisciplinary and emerging fields. The Millennium Project goes even further and demands a knowledge system that ensures sensitivity of policy-making to global long-term perspectives via the establishment of collective intelligence systems that provides continuity from one administration to the next and assists to cope with accelerating knowledge expansions, complexities, and interdependencies while securing public agreement about necessary changes [42]. Glenn and coworkers state that one of the major problems in policymaking are incoherent policies across countries. A related view from the online resources (http://www.millennium-project.org/millennium/Global_Challenges/chall-05.html) of the Millennium Project is: "Leaders should make these new systems as transparent and participatory as possible to include and increase the public's intelligence and resilience. As a result, more future-oriented and global-minded voters might elect leaders who are sensitive to global long-term perspectives. … Universities should fund the convergence of disciplines, teach futures research and synthesis as well as analysis, and produce generalists in addition to specialists." A tree of knowledge, that is accessible by all, and where connections and interdependencies can easily be visualized, would be a first approach to this set of challenges. Gebeshuber and Majlis proposed such an approach in their publication on new ways of scientific publishing and accessing human knowledge inspired by transdisciplinary approaches in tribology [34]. This would help researchers, policy makers and the general public to connect, collaborate, and communicate in order to facilitate the sharing of their knowledge.

Tribology funding bodies and policies increasingly incorporate ecotribology thought in their approaches, yielding promising prospects departing from singular monetary considerations toward responsible actions for the current and future generations.

3.5.1 Three Gaps Toward Sustainability

Inspired by the three-gaps theory that our group introduced in 2009 for inventors, innovators, and investors [35], a three-gaps theory for the path of tribology toward sustainability is proposed (Fig. 3). Successful bridging of these three gaps provides a potential path for successful ecotribology, contributing to propel society toward sustainability.

The three-gaps theory for ecotribology summarizes issues that need to be identified and addressed successfully for the establishment of successful ecotribology. The first gap is between the world of ideas and conventional tribology. Generally, innovation, disruptive ideas, and people who realize the potential of ideas overcome this gap (which acts as *push* in a push–pull theory approach, [96]). In conventional tribology, markets, individuals, and know-how are important. The second gap on the path toward successful ecotribology is to leave the focus on markets, and realize and appreciate the issues that arise for society due to global challenges. Instead of *know-how*, *know-why* becomes important. The effects of tribology are viewed with a broader scope, especially regarding their potential (in combination with other approaches) to successfully address our most pressing problems. The third gap is between identifying global challenges and realizing that also ecotribology is necessary to successfully address some of these challenges and subsequent development and implementation of successful ecotribology. In this way, needs of all life (not just selected human individuals, or countries, or societies, or corporations) are fulfilled and are the main issues of focus. Successful ecotribology is based on profound understanding. Various problems and issues need to be overcome when bridging this gap.

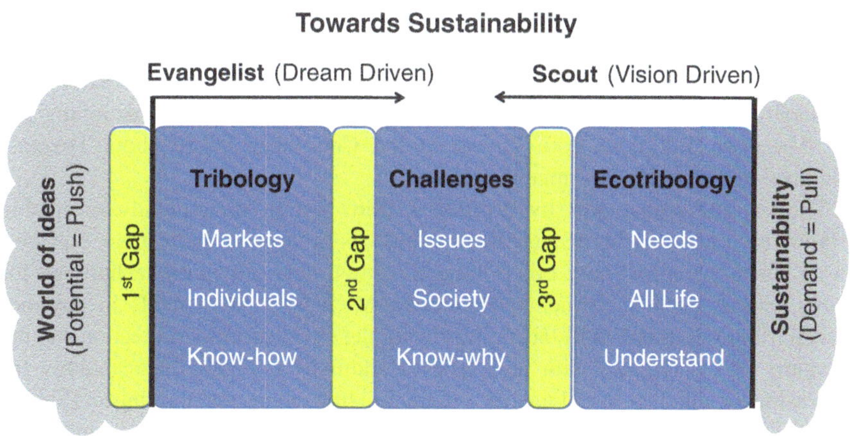

Fig. 3 Three-gaps theory for ecotribology. To propel tribology toward sustainability, three gaps need to be overcome: the one between the world of ideas and conventional tribology, the one between conventional tribology (practice, education) and the most pressing problems we are currently facing, as human societies and as all life of Earth, and the gap between the global challenges and ecotribology that is beneficial for the whole biosphere

Two specific personalities may arise in such a scenario: dream-driven *evangelists* and vision-driven *scouts*. The idea of scouts and evangelists comes from the classic 2007 paper by Hansen and Birkinshaw in the Harvard Business Review [49].

Ecotribology might be more expensive than conventional tribology when viewed with a narrow, economically oriented focus. In a wider focus, it is necessary—the ultimate goal is sustainability.

3.6 Current Efforts in Ecotribology Exemplified by the World Tribology Congress

International conferences are important for various reasons. First, new trends and developments are treated, and researchers have the possibility to listen to visionaries from their field and related fields, widening their horizons, thoughts, and fields of interest. Activities at conferences, such as personal discussion, entering scientific and technological networks, establishing contacts with industry and researchers alike pave the way for an even more successful career for young and established tribologists.

The World Tribology Congress takes place every 4 years. In 2013 it was in Italy while in 2009 it was in Japan. Ecotribology tracks have been organized at both recent World Tribology Congress events, in 2009 (main organizer: Prof. Shinya Sasaki from Tokyo University of Science, Japan) and 2013 (main organizers: Prof. Ille C. Gebeshuber, Austrian Center of Competence for Tribology, National University of Malaysia and Vienna University of Technology, and Prof. Jianbin Luo, Tsinghua University, China, and Director of the State Key Laboratory of Tribology).

Two plenary talks at the Ecotribology Track of the World Tribology Congress 2009 set the stage for the new and exciting field: Mr. Christopher Flaivin, the president of the Worldwatch Institute, USA, talked about low-carbon economy and Mr. Masatami Takimoto from Toyota Motor. Co., Japan talked about Toyota's initiatives for realizing sustainable mobility.

The keynote speech was by Wilfried J. Bartz, on environmentally acceptable tribological practices. Invited speakers talked about additive performance in bio-based versus paraffinic base oils (Girma Biresaw), new ecological technology for heavy-loaded machine elements (Remigiusz Michalczewski), trends in environmental tribology (Stephen M. Hsu), improving the environmental protection and the economy of i.c. engines with new type of additive (J. Fodor), ecotribology for increasing the efficient use of energy and minimizing environmental impact (Shinya Sasaki), the development of high performance shaft seal "leaf seal" in industrial turbines (Hidekazu Uehara), mixed lubrication analysis of vane sliding surface in rotary compressor mechanisms—influences of flexible structure at surface end of vane slot (Yasutaka Ito) and development of a human-friendly, renewable resource-based metalworking fluid technology, and its impact on sustainable manufacturing (Jake W. Pajak). Interesting to note is the balance between speakers from the industry and speakers from universities and research institutions and the broad approach to the field—this provides a very good basis for something that still needs to be defined properly, such as the new and emerging field of ecotribology.

The Ecotribology Track at the WTC2013 in Torino had five sessions: tribology for energy conservation (with the keynotes on green tribology at sea at Robert J.K. Wood and MEMS/NEMS and BioMEMS/BioNEMS materials and devices and biomimetics by Bharat Bhushan), environmentally friendly tribology, environmentally compatible lubricants, tribology for life—green tribology (with a surprise talk on sustainability by Prof. Satish V. Kailas from the Indian Institute of Science, the premier Indian tertiary education institution), and a panel discussion on ecotribology with an impulse talk [37] on ecotribology development, prospects, and challenges by the author of this chapter. Jianbin Luo, Zygmunt Rymuza, Ali Erdemir, Shi-wei Zang (who coined the term "Green Tribology"), Ille C. Gebeshuber—the author of this chapter, and the artist and ecologist Sigrid Zobl comprised the panel, moderated by Oliver Futterknecht. The session highlighted the importance of ecotribology across fields (ecology, tribology, economy, policy makers, etc.) and of incentives, also from the political side.

4 Sustainable Ecotribological Systems: Best Practice from Living Nature

This part of the chapter deals with inspiration from living nature for the development of successful ecotribology. Rationale for looking at organisms and ecosystems for inspiration is the following: organisms need to cope with issues related to friction, adhesion, lubrication, and wear. They have developed a sustainable system integrating various interesting approaches to tribological problems, such as lubrication of the eye and the hip, protection of inner linings in pregnant women, optimization of skin properties on finger tips (people can hold onto things, even with wet hands), and many more. In the subsections of this section, the inspiring aspects from living nature are sorted related to materials (exemplified by metals), structures (exemplified by structural colors with optimized tribological properties), and processes (exemplified by biomineralized structures with optimized tribological properties) and transferred to ecotribology.

4.1 Materials

Traditionally, metals are of high importance in engineering. In living nature, this is not the case. Apart from biomineralized structures (see Sect. 4.3) organisms only use metals when they are chemically necessary, for example, as the center atom of chlorophyll, Magnesium, and as the center atom of hemoglobin, Iron. Mechanical strength, structural support, and further functional properties in organisms are provided rather by highly functional structures of benign materials than by metals. Until mankind is with its technology at such a high stage, metals will still be of utmost importance in tribology.

Heavy metals in the soil and in wastewater pose a threat to human life and cause significant environmental problems. Some microorganisms and plants, however, have the ability to accumulate high amounts of metals in their bodies, some in such high amounts that they can be used for mining (biomining = mining with organisms). Biomining denotes the use of microorganisms and plants (phytomining, see Sect. 4.1.2) to aid the extraction and recovery of metals from ores and further metal-including materials such as electronics waste. The rare and specific cases of metal-accumulating organisms may serve here as inspiration for the potential development of an alternative, less environmentally damaging way to mine for metals for applications in tribology. This section investigates plants and microorganisms that accumulate metals, and that might be inspiration for alternative ways to obtain metallic resources.

Biomining with plants, phytomining, is less profitable than biomining with bacteria. However, this technology has found its niches, and especially in the production of nanoparticles with controlled shape and size, plants are of increasing interest for research and development [57]. Biomimetic resource management via mining with organisms provides innovative ways of interpreting waste, waste effluents, and pollutants as raw materials for research and industry, inspired by materials, structures, and processes in living nature [55].

4.1.1 Biomining with Bacteria

Mining with bacteria is arguably more environmentally friendly, more efficient, and cheaper than conventional mining [55, 70]. In conventional mining, metals are melted from ores in blast furnaces; poisonous sulfur gases, fine dust, and greenhouse gases need to be dealt with. In biomining with bacteria, the rocks and ores are finely grounded and treated with sulfuric acid (the rock-eating bacteria need acidic conditions). This allows the bacteria that are already in the material to grow at fast pace, and to enzymatically alter the oxidation state of the metals, gaining metabolic energy. Rock-eating bacteria feed by decomposing ores, metal containing waste, heavy metal loaded industrial effluents and finely ground mining waste, and rocks with metal content too low for conventional mining, leaving acidic fluids with high metal content from which the metals are subsequently gained via electrical processes.

Biomining with bacteria is well established in various countries: in South Africa, Canada, Australia, Chile, India, and China about 25 % of the Copper and 10 % of the Gold are produced with this method (status: 2015). Further metals that are mined with bacteria comprise Zinc, Nickel, Cobalt, Silver, Uranium, Indium, Germanium, Molybdenum, Palladium, and Lead.

4.1.2 Mining with Plants for Bulk Metals

Contaminated soil and contaminated wastewaters are the result of common manufacturing processes around the world. Plants (such as the sunflower plant [59], Fig. 4) and microorganisms such as bacteria, fungi, algae, and yeast can accumulate heavy metals and safely remove pollutants from waterbodies and the soil. In bioremediation approaches a contaminated environment is biotransformed back to its original pristine condition, with the help of organisms. High accumulation capacity can even be used for the enrichment or recycling of valuable metals. Bioremediation with plants involves minimal site destruction and is aesthetically favorable. Some plants even hyperaccumulate metals [4], i.e., they accumulate extraordinary high amounts of metals that are far in excess of the levels found in the majority of other species growing in the same location, without suffering phytotoxic effects.

Various plants mine the soil and the waterbodies in ways that are contrary to conventional human mining approaches. With their roots they take up metals from the ground, and accumulate them in their bodies. In some cases, various percents of the dry body mass of the plants are metal. When the metal is extracted and concentrated in the plant tissue, the plants are harvested, dried, ground, and burnt as

Fig. 4 The sunflower plant *Helianthus annuus* can be used in a new way of mining metals: phytomining—mining with plants. Photo by Kurt Jansson. Licensed under the creative commons attribution-share alike 2.0 Germany license. Image reproduced with permission

part of the metal extraction process (yielding also thermal energy). After the plant or biomass is burnt to produce bio-ore (small volume plant ash that contains the target metal in high concentrations), it is treated with chemicals to get refined metals. Learning from plants how to mine could potentially revolutionize our way to obtain base materials for our technological devices. In the future, however, a nearly complete replacement of metals by functional structures made from benign materials can be envisaged.

Protection from herbivore attack through feeding deterrence (the plants would taste bad) and from pathogen attack through their toxicity might be the reasons why plants hyperaccumulate metals.

Reeves [72] reports 440 plant hyperaccumulators, 75 % of which hyperaccumulate Nickel. The main characteristics of hyperaccumulators are metal accumulation, metal translocation from roots to shoots, metal enrichment, and metal tolerance.

The Yellowtuft *Alyssum murale* can hyperaccumulate Nickel up to 20 g/kg dry mass, and can reach a biomass of 10 tons per hectare. The Indian Mustard *Brassica juncea* can hyperaccumulate Gold via induced hyperaccumulation, [5] up to 57 mg/kg. Normal Gold concentration in plants is 0.01 mg/kg. Further commonly known hyperaccumulator plants are the Alpine Pennycress *Thlaspi caerulescens* (for Zinc) and the Copper Flowers *Haumaniastrum robertii* and *H. katangense* (for Copper) (Table 4).

Table 4 Hyperaccumulators used in phytomining (selection from 440 species)

Metal	Plant species	Metal concentration (mg/kg d.w.)	Biomass (t/ha)
Cd	Alpine pennycress (*Thlaspi caerulescens caerulescens*)	3000	4
Co	Copper flower (*Haumaniastrum robertii*)	10,200	4
Cu	Copper flower (*Haumaniastrum katangense*)	8356	5
Au[a]	Indian mustard (*Brassica juncea*)	10	20
Pb	Round-leaved pennycress (*Thlaspi rotundifolium*)	8200	4
Mn	*Virotia neurophylla*, flowering plant—IUCN status vulnerable	55,000	30
Ni	*Alyssum bertolonii*, alison species from Europe	13,400	9
	Berkheya coddii, a flowering plant from the daisy family	17,000	22
Tl	Buckler mustard (*Biscutella laevigata*)	4055	4
Zn	Alpine pennycress (*Thlaspi caerulescens calaminare*)	10,000	4

[a]Induced hyperaccumulation

4.1.3 Mining with Plants for Metallic Nanocrystals

People have argued that phytomining is not as interesting as biomining with bacteria, since the harvest is too small. With the arrival of the age of nanotechnology, and the importance of nanoparticles of specific shapes and sizes, plants and plant-based methods are increasingly important in yielding nanoparticles for research, development, and the market. The price of Gold nanoparticles (with diameters varying between 50–100 nm in one batch) is about 10 times higher than that for Gold in bulk; the price of Gold NPs rises to several thousands of dollars per gram the more defined the size and shape of the Gold nanoparticles need to be. Because of these economic reasons, NP production with plants is viewed with increasing interest.

Biosynthesis of inorganic nanoparticles comprises metallic nanoparticles (NPs), oxide NPs, sulfide NPs, and other typical NPs [57]. Iron oxide nanoparticles (Fe_3O_4-NPs) can be synthesized with Sargassum Algae (*Sargassum muticum*) aqueous extract, Gold nanoparticles are synthesized by the thermophilic bacterium *Geobacillus* sp., the fungus *Verticillium luteoalbum*, and the bacterium *Klebsiella pneumoniae*. The *Lactobacillus* from our yogurts biosynthesizes Silver and Titanium dioxide NPs. The bacterium *E. coli* and the black mold *Aspergillus niger* can produce Silver NPs.

Biosynthesis of Gold, Silver, Platinum, Palladium, silica, alloy, Titanium, zirconia, Selenium, and Tellurium nanoparticles by microbes has already been reported [63]. Also plants extracts such as fruit peel preparations can be used for the plant-mediated biosynthesis of nanoparticles via recycling of metal industrial effluents [57]. See Table 5 for some plant-extract produced NPs.

4.2 *Structures*

In Sect. 4.1, the way that plants and organisms mine for metals was introduced; a way that is so different to how we are currently doing it with conventional technology. Obtaining metals is important, especially at our current state of affair,

Table 5 Metal nanoparticles produced by plant extracts (selection)

NP material	NP size	Biosynthesizer
Ag	4.6 ± 2 nm	Pu-erh tea (*Camellia sinensis*) leaves [60]
	136 ± 10.09	Oregano (*Origanum vulgare*) leaves [76]
Au	10–30 nm	Damask rose (*Rosa damascena*) flowers [41]
	50 nm mean size	Betel (*Piper betle*) leaves [83]
	25 nm nanorods, 30 nm nanowires	Sugar beet (*Beta vulgaris*) pulp [14]
Pt	2.4 ± 0.8 nm	Oriental Arborvitae Leafytwigs (*Cacumen platycladi*) [95]
Se	60–80 nm	Lemon (*Citrus* sp.) leaves [69]

because various devices and technologies heavily rely on metals for their function. In the course of increasing sustainability of our technologies, however, metals and plastics made from fossil fuels need with time to be replaced with other, more benign materials.

Living nature is facing the same physical problems as our machines and devices. Constructions and devices with moving parts need to cope with gravity, weather conditions, loads, resist wear and tear, and fulfill further tribological requirements. One of the large differences between man-made and natural constructions is the type of materials used and the exquisite tailoring of structures in living nature, often on many levels of hierarchy [23]. The strong relationship between structure and function in organisms is one of the key reasons for the success of biomimetics.

Nachtigall identified general biomimetic principles that can be applied by engineers who are not at all involved in biology [62]. One of these principles is *integration instead of additive construction*. As opposed to the plenitude of different materials currently used in technological approaches, organisms use only a small amount of different materials, however, slightly chemically varied in different applications, and greatly varied structurally.

One example for the realization of the biomimetic principle *structure rather than material* in organisms is the way that colors are produced in some microorganisms and tropical plants [17, 36] as well as in animals such as the peacock, the pigeon, and fish. Minuscule structures on the order of the wavelength of the visible light (in some cases, of the order of the light visible to other animals, such as the ultraviolet light visible for some insects and birds) interact with the light. They do this in similar ways as the tiny water droplets in the air do in the generation of the colors of a rainbow, or a thin film of oil does on water when generating iridescent colors, i.e., colors that change with the viewing angle. The resulting brilliant structural coloration does not bleach, and—in most conditions—does not employ any potentially toxic chemicals but just contains structures from benign materials.

A central aspect of structural colors and structure-based functions in organisms in general that makes them interesting for tribology is their multifunctionality. *Multifunctionality instead of monofunctionality* is a further general biomimetic principle identified by Nachtigall [62]. Structures in and on some beetles, butterfly wings, tropical plants, and microorganisms yield impressive coloration, and also control wetting behavior as well as frictional and adhesion properties [65]. Examples for elaborately structured parts with tribological relevance are the nanostructured hydrated silica hinges and interlocking devices in diatoms (Fig. 5, [31]) that can serve as inspiration for optimized tribology in microelectromechanical systems (MEMS). On an even smaller scale, chemical ice binding properties based on specifically structured molecular adhesives in diatoms come to mind [71].

Diatoms are single-celled algae that biomineralize hydrated silica parts of amazing variety, forms, structures, and functions. There are tens of thousands of different diatom species, all with different morphology. Diatoms range in size from a few micrometers to a couple of millimeters, and some species build chain-like colonies. In the colonies the single cells are mechanically attached to each other by various mechanisms, from adhesives [91] to hinges [31] and nanozippers [87],

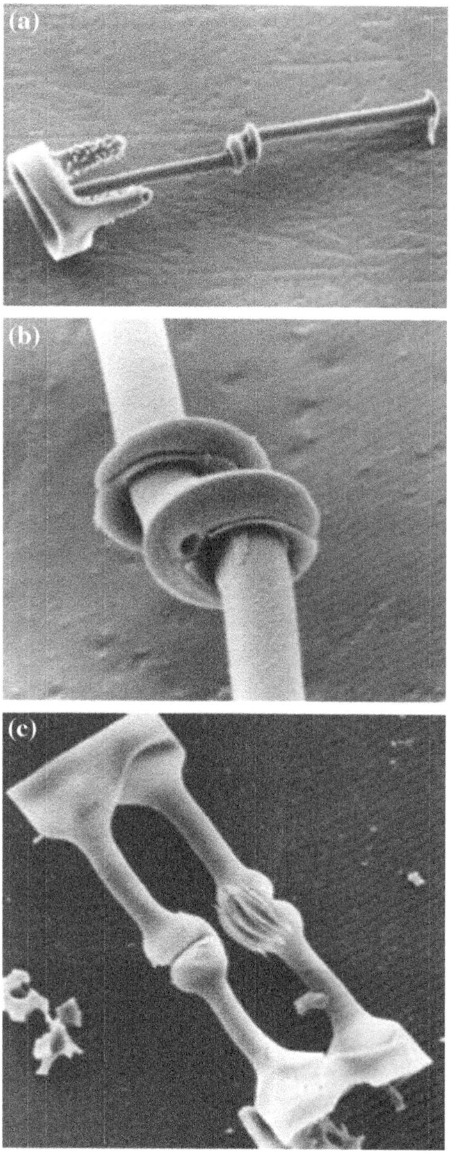

Fig. 5 Diatoms are single-celled algae that biomineralize hydrated silica parts of amazing variety, forms, structures, and functions, interesting for micro- and nanotribologists regarding their optimized tribological properties on the micro- and nanoscale. **a** In the fossil diatom genus *Syndetocystis*, the linking occurs by means of a structure at the center of the valve. One of the diatom valves has been broken and lost. At the apex of each spine is a loop that surrounds the shaft of the other spine. Scale bar 20 μm. **b** A detail of the apex is shown. Scale bar 5 μm. **c** Sibling valves of the diatom *Briggera* with a massive linking structure at either end of the oval valve. Scale bar 20 μm **a** and **b** from [31], **c** from [16]. Permission pending

curiously often permitting movement between cells without detachment [91]. There are several properties of the diatoms that make them interesting regarding MEMS and micro- and nanotribology [25, 38]. For example, they build their exoskeletons during cell division, and do not change the basic frustule after this. Repair does not take place on the hard silica parts. They use hydrated silica as material, and therefore accommodate various functions via structure alone. They are nanostructured, and their optical properties are interesting for nanotechnologists.

4.2.1 Structure Rather Than Material

In this section we give four specific, tribologically interesting examples for the realization of the biomimetic deep principle *structure rather than material* in organisms. Snake scales provide friction reduction; wear protection in iridescent red algae results as by-product (with potentially further functionalities, given the omnipresent multifunctional aspects of biomaterials) in strikingly beautiful iridescent coloration. The tropical butterfly *Morpho* has multifunctional iridescent blue wings. The author encountered *Morpho* butterflies in the virgin rainforest of Costa Rica, when she was doing biomimetics with engineers from the aircraft company, Boeing. The structural black coloration of *Troides* sp., the iridescent coloration of certain Malaysian tropical understory plants, and the calcite tips of marine microorganisms that are exquisitely tailored regarding growth along specific crystallographic axes are further examples from living nature on realizations of the deep principle *structure rather than material* with relevance for ecotribology (Table 6).

Snake scales for friction reduction

Iridescent snake scales might serve as friction reduction structures [24]. Biomimetic inspiration for frictionally optimized structures of household and other appliances might result in beautiful, functional iridescent surfaces. The brilliant strong coloration, especially when limited in bandwidth (i.e., showing over a wide viewing angle only one color, rather than all the colors of the rainbow, see scientific literature on the related structures in the scales of the *Morpho* butterfly and references therein) would be an additional bonus for luxury functional materials.

Wear protection in iridescent algae

The iridescent cuticles of the algae *Iridaea flaccida* and *I. cordata* are tough but flexible outer coverings. They were investigated by Transmission Electron Microscopy [39], and showed multilamination with alternating electron opaque and translucent layers with a total thickness of 0.5–1.6 μm. According to the publication, the electron opaque layers may correspond to protein-rich regions and the electron translucent ones to regions rich in carbohydrates. The cuticle may protect the alga from physical factors such as desiccation, wear, and from predator injury. The authors conclude that it is likely that the iridescence in other foliaceous red algae is caused by a similar structure.

Light and temperature regulation management in black butterflies

Herman and coworkers report in 2011 temperature regulation properties of the black wings of certain butterflies [50]. Butterfly wings need to have a certain temperature so that the animal can fly, but this temperature cannot be too high, because otherwise the proteins and other fragile biomaterials of the organism would disintegrate. Understanding of the contribution of the nanostructures of the black butterfly wings might provide valuable biomimetic inspiration for the development of tailored nanostructured technological materials that would express a passively controlled temperature range. Such materials would find wide use, from building skins to clothing to car surface treatment. The fact that the temperature regulation might substantially originate from the structure rather than the material would allow for property transfer to materials that are already used for the selected applications,

Table 6 Physical mechanisms that yield iridescence in plants, with examples and functions [17]

Physical mechanism	Visual appearance	Example	Function
Thin film interference	Multicolored pointillistic perdium	Slime mold *Diachea leucopoda*	Photoprotection
Multilayer interference	Iridescent blue leaves	Willdenow's spikemoss *Selaginella willdenowii*	Photoprotection
	Iridescent blue leaves	Peacock begonia *Begonia pavonina*	Photoprotection
	Rainbow colored blades	Red alga *Iridaea flaccida*	Wear protection
Diffraction gratings	Blue, green, and yellow iridescent flowers	Flower of an hour *Hibiscus trionum*	Attraction of pollinators
	Flowers iridescent in the UV range (visible for certain insects)	Tulip *Tulipa kolpakowskiana*	Attraction of pollinators
Scattering	Bluish white needles	Blue spruce *Picea pungens*	Preferential scattering of short wavelengths and enhanced reflectance of UV
Photonic crystals	Iridescent blue fruit	Blue quandong *Elaeocarpus angustifolius* syn. *E. grandis*	High visibility in green foliage, contribution to photosynthesis
	White hairs	Edelweiss *Leontopodium nivale* subsp. *alpinum*	UV protection, light guide
	Iridescent metallic blue fruit	*Pollia condensata*	Display and defense
Cholesteric liquid crystals	Iridescence in juvenile leaves	Fern *Danaea nodosa*	Photoprotection

and even allow for transfer to novel, potentially benign materials, for the applications. Imagine regulating the temperature inside cars, inside houses, and inside cooling containers (and containers that keep things warm) simply by passive structures! The manufacturing of the structures might be not cheap, but there would be no maintenance costs, rendering the total costs lower than the conventional, in many cases polluting technology. Controlling the temperature range for certain applications in a certain range would also allow for more specific tailoring of adhesive, lubricants, and additives, excluding the need to develop such tribologically optimized materials for extreme conditions.

Excellent structure-based multifunction of *Morpho* butterfly wings

Morpho butterfly wing scales have received enormous attention from the engineering and biomimetics communities (for review, see [65]). Famous for its iridescent wing scales due to periodic micro- and nanostructures of chitin and air layers, various functions have been identified. Tribologically interesting functions comprise thermal response, selective vapors response due to polarity gradients on the nanoscale (from more nonpolar close to the wing to more polar orthogonally away from the wing, [68]), superhydrophobicity, directional adhesion, and self-cleaning properties.

There are various *Morpho* butterflies, most of which show brilliant blue iridescence. The structures responsible for the multifunctional properties of the butterfly wings vary between species, between male and female butterflies, and with the location of the respective scales on the wings. Christmas-tree like periodic nanostructures act as multilayer reflectors and diffraction gratings, and in combination with ridges in the scales result in wide-angle appearance of the blue coloration. Direction adhesion provides water runoff properties away from the body, and allows the butterfly to fly in rain, without getting its body wet. The contact angle for water is above 150°.

Iridescent blue plants

Structural coloration in animals, plants, and microorganisms mainly appears on the surface of the respective organism. Biological surfaces are the first interfaces in the interaction process between the respective organism and its surrounding environment—another fact that makes them highly interesting for knowledge transfer to tribology.

The iridescent blue lycophyte Willdenow's Spikemoss, *Selaginella willdenowii*, is a common plant in the Malaysian rainforest [58]. It exhibits blue iridescence (Fig. 6; investigation with TEM reveals various layers with less than 100 nm thickness each) on the surface [58]. Figure 6 was taken in the Bukit Wang Recreational Forest in Malaysia, with very long exposure time as to better reveal the blue coloration of the leaves. Seen with the naked eye, the fern is bluish green and seems to glow in semidarkness. The blue coloration is iridescent and changes with the viewing angle. Holding and tilting the leaves results in color change from green to nearly total blue, although not as strong as seen in the long exposure time photograph shown in Fig. 6. It remains for future research to establish the exact functions of this strikingly blue coloration, and its potential inspirational potential for tribology.

Crystal spines in Coccolithophores

Coccolithophores are unicellular planktonic algae with a size of a couple of micrometers. Coccolithophores biomineralize an exoskeleton made of calcitic micro- and nanostructures, with exquisite control of shape, size, and crystal orientation (Fig. 7). Calcite is only one of more than 70 metals and alloys, ceramics, polymers, and composites that are biomineralized by organisms at ambient conditions. See Gebeshuber [28] for a review of biominerals and marine organisms that produce them, including various extensive lists of the materials that are biomineralized, the proteins that aid in biomineralization, and the potential functions of biomineralized structures, including prospects in engineering and medicine.

Fig. 6 The iridescent blue lycophyte *Selaginella willdenowii*, a fern ally, is a common plant in the Malaysian rainforest. © Mr. Foozi Saad, IPGM, Malaysia. Image reproduced with permission

The ultrastructure and crystallography of the coccolithophore *Rhabdosphaera clavigera* heterococcoliths was characterized by electron microscopy techniques, including three-dimensional electron tomography, by Young and Henriksen [93]. Five spiral staircases and discrete tip elements made from single crystal {104} calcite rhombohedra units are depicted in Fig. 7. Within the spine core, crystal platelets are hypothesized to serve as a template for nucleation and assembly of the overall structure.

Surface textures are currently an important field in tribology. From the biomineralization in coccolithophores and related organisms novel approaches to textured surfaces will gain inspiration regarding not just the texturing of surfaces of various length scales with potential hierarchical effects, see [23], but also additional synergistic benefit from the control of the shape, size, and crystallographic orientation of the structuring elements.

4.2.2 Tribologically Optimized Microstructures in Organisms

Above, various tribologically interesting microstructures in organisms are presented, and their inspirational potential for ecotribology is outlined. In their classic

Fig. 7 Spine tip of the microorganism *Rhabdosphaera clavigera*. Material: Calcite, $CaCO_3$. Structure: spiral formed from consistently aligned crystal units with rhombic faces. Process: Biomineralized in cool seawater, with the help of proteins, at ambient conditions. Scale bar 1 µm = 0.001 mm. Image Source: Young and Henriksen [93]. Image reproduced with kind permission of the Mineralogical Society of America

book, Biological micro- and nanotribology: Nature's Solutions, Scherge and Gorb [79] give various further examples. The fast technological progress in nanoscale characterization and production of functional micro- and nanostructures in recent years suggest revisiting the classic book, and developing interesting, potentially disruptive solutions for ecotribology.

4.2.3 Tribologically Optimized Nanostructures in Organisms

Research results such as the finding of a polarity gradient in the single Christmas-tree like nanostructures of the *Morpho* butterfly [68], as well as further nanoscale gradient properties that have been identified in organisms, such as the nanoscale material gradient in adhesive setae of a ladybird beetle [67] promise various additional research findings on such structures in organisms, and increased knowledge about their functionalities and technological ways to produce them, for optimization of tribological properties.

4.2.4 Bridging the Gap: Functionality Across Scales

Micro- and nanotribological investigations are important to establish basic tribological understanding. For transfer of the found principles, processes, etc. to applied tribology, the gaps between the small scales and the macroscale need to be bridged [88]. Bridging the gap between length scales is one specific issue in biomimetics, since more often than not organisms exhibit hierarchical properties, with functionalities at each length scale, and added, synergistic functionalities because of hierarchies [23].

In Sects. 4.1 and 4.2, materials and structures in organisms and potential knowledge transfer to tribology were introduced.

Many of the presented structures are orders of magnitude away from the macroscale: the length scale of their functional units spreads from the micro- and nanoregime to the macroscale: some *Morpho* butterflies have 20 cm wingspan. The different questions asked in the respective studies require scientific investigations at different resolutions. Scale effects need to be addressed when establishing tribological models across scales [12].

Microtribological investigations yield important information concerning wear, surface fractures, the formation of structures between microscopic parts of the tribosystem and their boundaries (tensions, shears, rupture, deformation, etc.). On the nanoscale, molecular properties can be probed. Depending on the questions to answer, investigations on the length scales of surfaces, clusters, and molecules might be necessary.

An interesting and challenging fact about tribology is that it is a systems science. The same holds for biomimetics. Detailed understanding of a whole tribosystem is dependent on understanding of the connections, interdependences, and single functionalities on all length scales of functionalities.

4.3 Processes

The third category in which ecotribology can learn from living nature is in the large and important area of processes. Materials and structures in living nature are produced via completely different approaches than materials and structures currently applied in tribosystems. One intriguing example for the exquisite processes employed in living nature is biomineralization of more than 70 different metals and alloys, ceramics, polymers, and composites by living organisms [61].

4.3.1 Biominerals

Such materials comprise carbonates such as $CaCO_3$ shells in mollusks, phosphates such as hydroxyapatite in bones, oxides such as magnetite Fe_3O_4 in bacteria, sulfates such as celestite $SrSO_4$ in radiolarians, sulfites such as pyrite FeS_2 and greigite

Fe_3S_4 in magnetotactic bacteria, arsenates such as oripiment As_2S_3 in bacteria, native elements such as Gold nanoparticles in yeast, silica $SiO_2 \cdot nH_2O$ in diatoms, halides such as fluorite CaF_2 in fish, and organic minerals such as guanine in fish scales, providing the beautiful fish silver (Table 7, [28]).

Biomineralization is characterized by chemical reactions involving proteins, the creation of perfect crystals, the control of crystal growth and inhibition depending on the crystallographic axis, as well as the production of composite materials with properties that are of high value to engineering.

One of the amazing properties of biomineralization in organisms is that material, structure, and function are strongly correlated.

The combination of minerals, which are usually stiff, brittle, and cheap energywise, with soft and pliable organic materials synergistically results in biominerals with amazing functionalities, such as bone with its seven levels of hierarchy.

4.3.2 Production of Microscopic Tribosystems in Freezing Water

Uncontrolled ice crystal formation inside living cells is dangerous for organisms. Various freeze protection proteins and ice nucleation proteins that promote and control freezing can be found in living cells. Technological applications of these materials are the production of artificial snow, antifreeze agents for aircraft and windshields, and novel, protein-assisted ways to control the feeling of frozen food in the mouth (cf. oral tribology). One example is the use of ice nucleation proteins in water-based ice cream, causing smooth and milky texture, comparable to the high calorie milk-based ice cream.

Various diatoms contain such proteins, and biomineralize their hydrated silica structures in saltwater at subzero temperatures. An intriguing example is the polar diatom *Corethron* [31]. *Corethron* is packed with tribologically interesting rigid parts in relative motion, such as a click-stop mechanism and chiral hooked spines, serving as inspiration for novel 3D MEMS.

4.3.3 Proteins Controlling Biomineralization

Scientists are just beginning to understand the role of proteins in biomineralization [92] and its tremendous potential for the tailor-made production of functional nano-, micro, meso-, and macrostructures with optimized tribological properties [30].

Active roles of proteins in biomineralization comprise inhibition of spontaneous mineral formation from solution (e.g., the protein statherin in the mouth inhibits spontaneous precipitation of Calcium phosphate), inhibition of growth of existing crystals, and responsibility for directing crystal nucleation, phase, morphology, and growth dynamics.

The shape of biomineralized crystals (such as the single crystal teeth of sea urchins) is affected by proteins with specific structures and sequences that adsorb to different faces of the crystal, leading to regulation of shape (crystal faces have

Table 7 Exemplary list of biominerals produced by organisms

Biomineral	Chemical formula	Biomineralizing organism
Carbonates		
Calcite	$CaCO_3$	Molluscs
Strontianite	$SrCO_3$	Snails, microbes, cyanobacteria
Siderite	$FeCO_3$	Bacteria
Phosphates		
Hydroxyapatite	$Ca_5[OH\|(PO_4)_3]$	Vertebrates (bones, teeth, fish scales), snails (grating tongue)
Whitlockite	$Ca_{18}H_2(Mg, Fe)^{2+}{}_2(PO_4)_{14}$	Dental plaque
Vivianite	$Fe^{3+}{}_2(PO_4)_2 \cdot 8H_2O$	Bacteria
Oxides and hydroxides		
Magnetite	Fe_3O_4	Eubacteria, archaeobacteria, teleosteans, chitons
Amorphous ilmenite	$Fe^{+2}TiO_3$	Foraminifera, snails (grating tongue)
Akaganeite	β-FeOOH	Bacteria
Sulfates		
Celestite	$SrSO_4$	Radiolarians, acantharia, algae, foraminifera, snails (shell)
Jarosite	$KFe^{3+}{}_3(SO_4)_2(OH)_6Z$	Fungi, bacteria
Barite	$BaSO_4$	Bacteria, algae, diatoms, foraminifera, loxedes (gravity receptor), clams
Sulfites		
Pyrite	FeS_2	Magnetotacic bacteria
Galena	PbS	Sulfate reducing bacteria
Greigite	Fe_3S_4	Magnetotactic bacteria
Arsenates		
Oripiment	As_2S_3	Bacteria
Native elements		
Gold	Au nanoparticles	Yeast, sponges, algae
Silver	Ag nanoparticles	Fungi
Uranium	U nanoparticles	Bacteria
Silicates		
Hydrated silica	$SiO_2 \cdot nH_2O$	Radiolarians, diatoms, glass sponges, limpets, mollusks
Halides		
Fluorite	CaF_2	Skeletons, fish skin, mollusk shells
Hieratite	K_2SiF_6	Mammalian gravity receptors
Atacamite	$Cu_2(OH)_3Cl$	Bloodworms
Organic minerals		
Guanine	$C_5H_5N_5O$	Fish, spiders

(continued)

Table 7 (continued)

Biomineral	Chemical formula	Biomineralizing organism
Magnesium oxalate (Glushinskite)	$Mg(C_2O_4)\cdot 2(H_2O)$	Plants
Copper oxalate (Moolooite)	$Cu(C_2O_4)\cdot 0.4H_2O$	Plants, fungi, mammals

Adapted from [28]

different charges and arrangements of atoms so proteins can selectively adsorb). Furthermore, proteins can self-assemble into ordered arrays that guide the formation of organized mineralized structures. Transferred to tribology, such protein-mediated approaches would allow for controlled, tailored growth of tribosystems at ambient conditions, with materials drawn from local resources, and little or no waste.

Common principles in the three main organic structuring and scaffolding polymers, chitin, cellulose, and collagen, are nanofibril formation (1.5–2 nm diameter), ability to self-assemble, production of fibrillar and fiberlike structures with hierarchical organization from the nanolevel up to macrolevels, ability to act as scaffolds and as templates for biomineralization, and the formation of rigid skeletal structures [22].

Certain proteins provide active organic matrices that control the formation of specific mineral structures; others act as catalysts that facilitate the crystallization of certain metal ions [92].

4.4 *Knowledge Transfer to Ecotribology*

4.4.1 Common Language

Ecotribology is a highly diverse field, with actors from a multitude of areas of specialization, working in basic or applied research, or in industry. It is increasingly harder to identify workforce with the diverse skills demanded from a tribologist.

Due to increased specialization in science and engineering, many people work just in one small aspect of their respective field—this also holds for tribologists. Such specialists increasingly get to know their area better and better but in some cases have no time or think they have no reason to talk to specialists of related but slightly different fields. In the extreme, the specialist languages become too detailed, no joint language can be reached across fields, and writings and oral presentations of authors only reach an audience coming from the same field.

If we are to understand biological tribosystems and develop theoretical models across scales that can fulfill the requirements we demand from models in science and technology, namely, to provide a way to predict the behavior and performance of other, unrelated systems, knowledge transfer between the fields of specialization needs to be ensured. Only in this way, good biomimetic technology is the result, and only properties that are intended to be transferred are indeed transferred.

Functionalities on the nanoscale influence functionalities on the microscale and subsequently on the macroscale. One of the goals in basic tribology research is a unified approach to energy-dissipating systems that encompasses most tribological phenomena.

Three needs can be identified regarding successful development of such a unified approach to energy-dissipating systems: We need a joint language, a joint way of publishing results and joint seminars, workshops, and conferences. Developing these three needs further results in a general concept concerning the future of scientific publications and ordering as well as accessing the knowledge of our time [34].

Currently, over-information in almost any field is a problem. Jack Sandweiss, editor of Physical Review Letters for 25 years, stated in 2009 in an editorial address "*For example, it is currently impossible for anyone to read all of Physical Review Letters or even to casually browse each issue*" [75]. Sandweiss refers to just one single journal!

Gebeshuber and Majlis suggested in 2009 dynamic publications of variable length that use various types of multimedia with adaptive information content. One and the same "paper" would be accessible to readers from various backgrounds and areas of specialization. In case more detailed information is needed, simple clicks on the links would expand the "paper" in the direction(s) wanted. "Recommendation agents" of the future could constrain information and thereby protect users from over-information by making the number of recommendations a function of the user's ability and readiness information intake.

In this way, ecotribology generalists would emerge; people who ensure knowledge transfer from one area of specialization to the other—contributing to a sound foundation to establish a unified approach to energy-dissipating systems across scales.

4.4.2 Biomimetics

Interacting surfaces in relative motion occur in tribology as well as in organisms. Biomimetics, the "Abstraction of good design from Nature" (definition from the Centre of Biomimetics at the University of Reading, UK) can therefore provide valuable developments in ecotribology. Specifically, the field of biomimetics deals with the identification of deep principles in living nature and their transfer to humankind [46]. The German biologist Werner Nachtigall identified ten general biological principles [62]. They can be applied by engineers who are not at all involved in biology, and are: integration instead of additive construction, optimization of the whole instead of maximization of a single component feature, multifunctionality instead of monofunctionality, fine-tuning regarding the environment, energy efficiency, direct and indirect usage of solar energy, limitation in time instead of unnecessary durability, full recycling instead of piling waste, interconnectedness as opposed to linearity and development via trial-and-error processes.

Currently, with nanotechnology as booming new promising field, the sciences have started to converge, with nanobioconvergence as one of the most promising examples. New materials, structures, and processes can arise from this novel approach, perhaps exhibiting some of the multifunctional properties of the inspiring organisms, where various levels of hierarchy (with additional functionality on each level) are so refined. We are hopeful that fully developed ecotribology will also exhibit one of the most important properties ensuring continuity of the biosphere: sustainability.

5 Conclusions and Outlook

Ecotribology is an exciting new field of tribology combining attention to ecological as well as economical aspects. Biomimetic approaches inspired by the oldest sustainable system we know, living nature, might pave the way toward efficient ecotribology, combining ecological and economical interests, and provide a liveable future for all.

Rich chemistry is found within interfaces in sliding contacts and there is a tight connection between chemistry and mechanics. Chemistry is of utmost importance in new ways to deal with resources, materials, structures, and functions. We need to foster the necessary skills to develop programmable materials, we need to slightly but effectively change base materials to obtain certain functions, and we need to establish a base set of materials that, in all sizes and shapes, are not harmful for people and the environment. Shape gets increasingly important, the smaller the particle—the toxicity of nanoparticles paramountly depends on shape, and shape on the nanoscale, can make a certain material either benign, neutral, dangerous, or highly toxic. One possibility to address the toxicity issues on the nanoscale might be to make dangerous and highly toxic nanomaterials fuse together, on demand, via a signal, resulting in harmless larger structures. This might be important for scenarios where nanomaterials leave reaction containers (inside they could very well be used for certain reactions, ensuring rendering them harmless when outside).

Career prospects in ecotribology are soaring, and contributing to the development of successful ecotribology is satisfying on the personal level as well as on societal levels—therefore, it is a highly recommended field for young tribologists to enter.

References

1. Agrawal A, Yun TH, Pesek SL, Chapman WG, Verduzco R (2014) Shape-responsive liquid crystal elastomer bilayers. Soft Matter 10:1411–1415
2. Allchin D (1998) Values in science and in science education. In: Fraser BJ, Tobin KG (eds) International handbook of science education. Kluwer, Dordrecht
3. Allchin D (2001) Values in science: an educational perspective. In: Bevilacqua F et al (eds) Science education and culture: the contribution of history and philosophy of science. Springer, New York
4. Anderson CWN (2013) Hyperaccumulation by plants. In: Hunt AJ (ed) Element recovery and sustainability. Royal Society of Chemistry, London
5. Anderson CWN, Brooks RR, Chiarucci A, LaCoste CJ, Leblanc M, Robinson BH, Simcock R, Stewart RB (1999) Phytomining for nickel, thallium and gold. J Geochem Expl 67(1–3):407–415
6. Anonymous (2010) Summary: world tribology congress 2009 (WTC IV) International tribology council information 191. http://www.itctribology.org/itcnews.php?issue=191&Go2=Go. Accessed 3 June 2015
7. Balazs AC, Aizenberg J (2014) Reconfigurable soft matter. Soft Matter 10:1244–1245. doi:10.1039/c4sm90006e
8. Barnosky AD, Hadly EA, Bascompte J, Berlow EL, Brown JH, Fortelius M, Getz WM, Harte J, Hastings A, Marquet PA, Martinez ND, Mooers A, Roopnarine P, Vermeij G, Williams JW, Gillespie R, Kitzes J, Marshall C, Matzke N, Mindell DP, Revilla E, Smith AB (2012) Approaching a state shift in Earth's biosphere. Nature 486:52–58
9. Barnosky AD, Matzke N, Tomiya S, Wogan GOU, Swartz B, Quental TB, Marshall C, McGuire JL, Lindsey EL, Maguire KC, Mersey B, Ferrer EA (2011) Has the Earth's sixth mass extinction already arrived? Nature 471:51–57
10. Bartz WJ (2006) Ecotribology: environmentally acceptable tribological practices. Tribol Int 39:728–733
11. Berne RW (2006) Nanotalk: conversations with scientists and engineers about ethics, meaning, and belief in the development of nanotechnology. Lawrence Erlbaum Associates, Mahwah
12. Bhushan B, Nosonovsky M (2008) Scale effects in mechanical properties and tribology. In: Bhushan B (ed) Nanotribology and nanomechanics—an introduction, 2nd edn. Springer, Berlin
13. Brooijmans RJW, Pastink MI, Siezen RJ (2009) Hydrocarbon-degrading bacteria: the oil-spill clean-up crew. Microbial Biotech 2(6):587–594
14. Castro L, Blázquez ML, González F, Muñoz JA, Ballester A (2010) Extracellular biosynthesis of gold nanoparticles using sugar beet pulp. Chem Eng J 164(1):92–97
15. Chaudhary K, Juarez JJ, Chen Q, Granick S, Lewis JA (2014) Reconfigurable assemblies of Janus rods in AC electric fields. Soft Matter 10:1320–1324
16. Crawford RM, Gebeshuber IC (2006) Harmony of beauty and expediency. Sci First Hand 5 (10):30–36
17. Diah SZM, Karman SB, Gebeshuber IC (2014) Nanostructural colouration in Malaysian plants: lessons for biomimetics and biomaterials. J Nanomat. doi:10.1155/2014/878409
18. Diamond J (2005) Collapse: how societies choose to fail or succeed. Viking Books, New York
19. Doucet SM, Meadows MG (2009) Iridescence: a functional perspective. J R Soc Interface 6: S115–S132
20. Drack M (2002) Bionik und Ecodesign—Untersuchung biogener Materialien im Hinblick auf Prinzipien, die für eine umweltgerechte Produktgestaltung nutzbar sind. Dissertation, Vienna University
21. Drack M, Gebeshuber IC (2013) Comment on "Innovation through imitation: biomimetic, bioinspired and biokleptic research" by A. E. Rawlings, J. P. Bramble and S. S. Staniland. Soft Matter, 2012, 8, 6675. Soft Matter 9:2338–2340

22. Ehrlich H (2010) Biological materials of marine origin: invertebrates. Biologically-Inspired Systems. Springer, Dortrecht Heidelberg London New York
23. Fratzl P, Weinkamer R (2007) Nature's hierarchical materials. Progr Mat Sci 52(8):1263–1334
24. Gans C, Baic D (1977) Regional specialization of reptile scale surfaces: relation of texture and biologic role. Science 195:1348–1350
25. Gebeshuber IC (2007) Biotribology inspires new technologies. Nano Today 2(5):30–37
26. Gebeshuber IC (2012) Green nanotribology and sustainable nanotribology in the frame of the global challenges for humankind. In: Nosonovsky M, Bhushan B (eds) Green tribology—biomimetics, energy conservation, and sustainability. Springer, Berlin
27. Gebeshuber IC (2012) Green nanotribology. Proc IMechE Part C J Mech Eng Sci 226 (C2):374–386
28. Gebeshuber IC (2015) Biomineralization in marine organisms. In: Kim S-K (ed) Springer hand-book of marine biotechnology. Springer, Tokyo
29. Gebeshuber IC (2015) Innovision in ecotribology: biomimetic approaches. In: Masjuki H et al (eds) Malaysian international tribology conference 2015. Penang
30. Gebeshuber IC (2015) Value based science: what we can learn from micro- and nanotribology. Tribology. doi:10.1179/1751584X15Y.0000000008
31. Gebeshuber IC, Crawford RM (2006) Micromechanics in biogenic hydrated silica: hinges and interlocking devices in diatoms. Proc IMechE Part J J Eng Tribol 220(J8):787–796
32. Gebeshuber IC, Macqueen MO (2013) New Asian case method for tribology: a structured approach for increased problem solving competence in tribology education and research. Tribology 7(2):69–73
33. Gebeshuber IC, Macqueen MO (2014) What is a physicist doing in the jungle? Biomimetics of the rainforest. Appl Mech Mat 461:152–162
34. Gebeshuber IC, Majlis BY (2010) New ways of scientific publishing and accessing human knowledge inspired by transdisciplinary approaches. Tribology 4(3):143–151
35. Gebeshuber IC, Gruber P, Drack M (2009) A gaze into the crystal ball—biomimetics in the year 2059. Proc IMechE Part C J Mech Eng Sci 223(C12):2899–2918
36. Gebeshuber IC, Lee DW (2016) Nanostructures for coloration (organisms other than animals). In: Bhushan B (ed) Springer encyclopedia of nanotechnology, 2nd edn. Springer, New York
37. Gebeshuber IC, Luo J, Prakash B, Rymuza Z (2013) Impulse talk: ecotribology—development, prospects and challenges. In: Ciulli E et al (ed) 5th World tribology congress WTC2013. Torino
38. Gebeshuber IC, Majlis BY, Stachelberger H (2011) Biomimetics in Tribology. In: Biomimetics—materials, structures and processes. Examples, ideas and case studies. Springer, Heidelberg
39. Gerwick WH, Lang NJ (1977) Structural, chemical and ecological studies on iridescence in Iridaea (Rhodophyta). J Phycol 13:121–127
40. Geryak R, Tsukruk VV (2014) Reconfigurable and actuating structures from soft materials. Soft Matter 10:1246–1263
41. Ghoreishi SM, Behpour M, Khayatkashani M (2011) Green synthesis of silver and gold nanoparticles using *Rosa damascena* and its primary application in electrochemistry. Physica E Low-Dim Sys Nanostruct 44(1):97–104
42. Glenn JC, Gordon TJ, Florescu E (2012) 2012 State of the future. MP Publications, Washington
43. Glenn JC, Gordon TJ, Florescu E (2014) 2013–14 State of the future. MP Publications, Washington
44. Glotzer SC, Solomon MJ (2007) Anisotropy of building blocks and their assembly into complex structures. Nat Mat 6:557–562
45. Godemann J (2008) Knowledge integration: a key challenge for transdisciplinary cooperation. Env Edu Res 14(6):625–641
46. Gruber P, Bruckner D, Hellmich C, Schmiedmayer H-B, Stachelberger H, Gebeshuber IC (2011) Biomimetics—materials, structures and processes. examples, ideas and case studies. In: Ascheron C (ed) Series: biological and medical physics, biomedical engineering. Springer,

Heidelberg Dordrecht London New York. p 266, 123 illus (53 in color), Hardcover, ISBN 978-3-642-11933-0, eISBN 978-3-642-11934-7

47. Guston DH (2006) Book review: a still small voice. Rosalyn W Berne 2006. Nanotalk: conversations with scientists and engineers about ethics, meaning, and belief in the development of nanotechnology (Lawrence Erlbaum Associates, Publishers: Mahwah, NJ). J Nanopart Res 8(1):149–152
48. Hamilton DP (1991) Research papers: who's uncited now? Science 251:25
49. Hansen MT, Birkinshaw J (2007) The innovation value chain. Harvard Bus Rev 85(6):121–133
50. Herman A, Vandenbem C, Deparis O, Simonis P, Vigneron JP (2011) Nanoarchitecture in the black wings of *Troides magellanus*: a natural case of absorption enhancement in photonic materials. Proc SPIE 8094, Nanophotonic Materials VIII. doi:10.1117/12.890946
51. Hinkel J (2008) Transdisciplinary knowledge integration. Cases from integrated assessment and vulnerability assessment. Dissertation, Wageningen University
52. Housman RF (1992) Sustainable living: Seeking instructions for the future: Indigenous peoples' traditions and environmental protection. Touro J Int Law 141(3):151–152
53. Jacobson A, Kammen DM (2005) Science and engineering research that values the planet. Bridge 35(4):11–17
54. Johnson RD (2012) Custom labware: Chemical creativity with 3D printing. Nat Chem 4:338–339
55. Karman SB, Diah SZM, Gebeshuber IC (2015) Raw materials synthesis from heavy metal industry effluents with bioremediation and phytomining: a bio-mimetic resource management approach. Adv Mat Sci Eng. doi:10.1155/2015/185071
56. Kramb RC, Buskohl PR, Slone C, Smith ML, Vaia RA (2014) Autonomic composite hydrogels by reactive printing: materials and oscillatory response. Soft Matter 10:1329–1336. doi:10.1039/c3sm51650d
57. Kulkarni N, Muddapur U (2014) Biosynthesis of metal nanoparticles: a review. J Nanotech. doi:10.1155/2014/510246
58. Lee D (2007) Nature's palette: the science of plant color. The University of Chicago Press, Chicago and London
59. Lin J, Jiang W, Liu D (2003) Accumulation of copper by roots, hypocotyls, cotyledons and leaves of sunflower (*Helianthus annuus* L.). Biores Tech 86(2):151–155
60. Loo YY, Chieng BW, Nishibuchi M, Radu S (2012) Synthesis of silver nanoparticles by using tea leaf extract from *Camellia sinensis*. Int J Nanomed 7:4263–4267
61. Mann S (2002) Biomineralization. Oxford University Press, Oxford
62. Nachtigall W (1997) Vorbild Natur: Bionik-Design für funktionelles Gestalten. Springer, Berlin
63. Narayanan KB, Sakthivel N (2010) Biological synthesis of metal nanoparticles by microbes. Adv Colloid Interf Sci 156(1–2):1–13
64. Nihongaki Y, Kawano F, Nakajima T, Sato M (2015) Photoactivatable CRISPR-Cas9 for optogenetic genome editing. Nat Biotechnol. doi:10.1038/nbt.3245
65. Niu S, Li B, Mu Z, Yang M, Zhang J, Han Z, Ren L (2015) Excellent structure-based multifunction of *Morpho* butterfly wings: a review. J Bionic Eng 12(2):170–189
66. Nosonovsky M, Bhushan B (2012) Green tribology—biomimetics, energy conservation, and sustainability. Series: green energy and technology, Springer Berlin Heidelberg
67. Peisker H, Michels J, Gorb SN (2013) Evidence for a material gradient in the adhesive tarsal setae of the ladybird beetle *Coccinella septempunctata*. Nat Commun. doi:10.1038/ncomms2576
68. Potyrailo RA, Starkey TA, Vukusic P, Ghiradella H, Vasudev M, Bunning T, Naik RR, Tang ZX, Larsen M, Deng T, Zhong S, Palacios M, Grande JC, Zorn G, Goddard G, Zalubovsky S (2013) Discovery of the surface polarity gradient on iridescent *Morpho* butterfly scales reveals a mechanism of their selective vapor response. Proc Natl Acad Sci 110:15567–15572

69. Prasad KS, Patel H, Patel T, Patel K, Selvaraj K (2013) Biosynthesis of Se nanoparticles and its effect on UV-induced DNA damage. Coll Surf B Biointerf 103:261–266
70. Rawlings DE, Johnson DB (2007) Biomining. Springer, Berlin
71. Raymond JA, Knight CA (2003) Ice binding, recrystallization inhibition, and cryoprotective properties of ice-active substances associated with Antarctic sea ice diatoms. Cryobiol 46:174–181
72. Reeves RD (2006) Hyperaccumulation of trace elements by plants. In: Morel JL, Echevarria G, Goncharova N (eds) Phytoremediation of metal-contaminated soils. Springer, New York
73. Rikken RSM, Nolte RJM, Maan JC, van Hest JCM, Wilson DA, Christianen PCM (2014) Manipulation of micro- and nanostructure motion with magnetic fields. Soft Matter 10:1295–1308
74. Rockström J, Steffen W, Noone K, Persson Å, Chapin FS III, Lambin EF, Lenton TM, Scheffer M, Folke C, Schellnhuber HJ, Nykvist B, de Wit CA, Hughes T, van der Leeuw S, Rodhe H, Sörlin S, Snyder PK, Costanza R, Svedin U, Falkenmark M, Karlberg L, Corell RW, Fabry VJ, Hansen J, Walker B, Liverman D, Richardson K, Crutzen P, Foley JA (2009) A safe op-erating space for humanity. Nature 461:472–475
75. Sandweiss J (2009) Essay: the future of scientific publishing. Phys Rev Lett. doi:10.1103/PhysRevLett.102.190001
76. Sankar R, Karthik A, Prabu A, Karthik S, Shivashangari KS, Ravikumar V (2013) *Origanum vulgare* mediated biosynthesis of silver nanoparticles for its antibacterial and anticancer activity. Coll Surf B Biointerf 108:80–84
77. Santoro AE, Lombardero MJ, Ayres MP, Ruel JJ (2001) Interactions between fire and bark beetles in an old growth pine forest. For Ecol Manage 144(1–3):245–254
78. Sasaki S (2010) Environmentally friendly tribology (Eco-tribology). J Mech Sci Technol 24 (1):67–71
79. Scherge M, Gorb SN (2001) Biological micro- and nanotribology: nature's solutions. Springer, Berlin
80. Scherge M, Dienwiebel M (2010) Book of synopses 17th int coll tribology: solving friction and wear problems, Technische Akademie Esslingen TAE. In: Bartz WJ (ed) Levers of tribological optimization. Ostfildern, p 13
81. Seckbach J, Gordon R (2016) Bio-Communication. World Scientific, Singapore
82. Smith ML, Lee KM, White TJ, Vai RA (2014) Design of polarization-dependent, flexural-torsional deformation in photo responsive liquid crystalline polymer networks. Soft Matter 10:1400–1410. doi:10.1039/c3sm51865e
83. Sneha K, Sathishkumar M, Kim S, Yun Y-S (2010) Counter ions and temperature incorporated tailoring of biogenic gold nanoparticles. Proc Biochem 45(9):1450–1458
84. Steffen W, Richardson K, Rockström J, Cornell SE, Fetzer I, Bennett EM, Biggs R, Carpenter SR, de Vries W, de Wit CA, Folke C, Gerten D, Heinke J, Mace GM, Persson LM, Ramanathan V, Reyers B, Sörlin S (2015) Planetary boundaries: guiding human development on a changing planet. Science 347(6223). doi:10.1126/science.1259855
85. Studart AR, Erb RM (2014) Bioinspired materials that self-shape through programmed microstructures. Soft Matter 10:1284–1294. doi:10.1039/c3sm51883c
86. Symes MD, Kitson PJ, Yan J, Richmond CJ, Cooper GJT, Bowman RW, Vilbrandt T, Cronin L (2012) Integrated 3D-printed reactionware for chemical synthesis and analysis. Nat Chem 4:349–354
87. Tiffany MA, Gordon R, Gebeshuber IC (2010) *Hyalodiscopsis plana*, a sublittoral centric marine diatom, and its potential for nanotechnology as a natural zipper-like nanoclasp. Polish Bot J 55:27–41
88. Tomala A, Goecerler H, Gebeshuber IC (2013) Bridging nano- and microtriblogy in mechanical and biomolecular layers. In: Bhushan B (ed) Scanning probe microscopy in nanoscience and nanotechnology III. Springer, Heidelberg
89. Tripathi A, Shum H, Balazs AC (2014) Fluid-driven motion of passive cilia enables the layer to expel sticky particles. Soft Matter 10:1416–1427

90. Urbakh M, Klafter J, Gourdon D, Israelachvili J (2004) The nonlinear nature of friction. Nature 430:525–528
91. Ussing AP, Gordon R, Ector L, Buczkó K, Desnitski A, VanLandingham SL (2005) The colonial diatom "*Bacillaria paradoxa*": chaotic gliding motility, Lindenmeyer model of colonial morphogenesis, and bibliography, with translation of OF Müller (1783), "About a peculiar being in the beach-water". Diatom Monogr 5:1–140
92. Wang L, Nilsen-Hamilton M (2013) Biomineralization proteins: from vertebrates to bacteria. Front Biol 8(2):234–246
93. Young JR, Henriksen K (2003) Biomineralization within vesicles: the calcite of coccoliths. Rev Min Geochem 54:189–215
94. Zang D, Lin K, Wang W, Gu Y, Zhang Y, Geng X, Binks BP (2014) Tunable shape transformation of freezing liquid water marbles. Soft Matter 10:1309–1314
95. Zheng B, Kong T, Jing X, Odoom-Wubah T, Li X, Sun D, Lu F, Zheng Y, Huang J, Li Q (2013) Plant-mediated synthesis of platinum nanoparticles and its bioreductive mechanism. J Colloid Interf Sci 396:138–145
96. Zmud RW (1984) An examination of "push-pull" theory applied to process innovation in knowledge work'. Manag Sci 30(6):727–738

Chapter 2
Advancements in Eco-friendly Lubricants for Tribological Applications: Past, Present, and Future

Carlton J. Reeves and Pradeep L. Menezes

Abstract This chapter highlights the evolution of eco-friendly lubricants derived from natural oils and fats to green lamellar solid additives to a new class of "greener" functional fluids known as room temperature ionic liquids (RTILs). The attraction to these bio-based lubricants began with vegetable oils due to their low friction and wear properties. These superior tribological characteristics are a result of their chemical composition of triacylglycerol molecules made up of esters derived from glycerol and long chains of polar fatty acids. It is these fatty acids within the natural oils that establish monolayers that enable high lubricity in boundary-lubricated regimes. Despite these accolades, vegetable oils suffer from thermal-oxidative instability, high pour points, and inconsistent chemical compositions. To improve upon the tribological properties, vegetable oils were subjected to additives such as lamellar solid powders to establish more resilient transfer layers to mitigate wear and surface damage. Currently, RTIL lubricants derived from bio-based feedstock represent a promising potential solution to many of the problems associated with previous eco-friendly lubricants. An investigation into RTILs begins with a discussion on the history of ionic liquids and an assessment on their tribological properties. The chapter also includes a case study on the use of RTILs as additives in vegetable oils and as neat lubricants as well as exploring the effects of cation-anion moiety exchange within ionic liquids themselves. Ultimately, the RTILs are compared to more traditional bio-based lubricants for their tribological performance as a new class of eco-friendly lubricants and their potential as a future lubrication technology.

C.J. Reeves
Department of Mechanical Engineering, University of Wisconsin-Milwaukee, Milwaukee, WI 53211, USA

C.J. Reeves · P.L. Menezes (✉)
Department of Mechanical Engineering, University of Nevada-Reno, Reno, NV 89557, USA
e-mail: pmenezes@unr.edu

J.P. Davim (ed.), *Ecotribology*, Materials Forming, Machining and Tribology,
DOI 10.1007/978-3-319-24007-7_2

1 Lubrication Fundamentals

A lubricant is a substance introduced between two moving surfaces to reduce friction, minimize wear, distribute heat, remove contaminants, and improve efficiency. The importance of lubricants and sustainable lubrication systems cannot be fully appreciated until understanding the implications of not using an appropriate lubricant or a lubricant at all. In 1979, it was estimated that over \$200 billion was spent in North America on machine maintenance [1]. Within the \$200 billion, approximately one-third (\$66.7 billion) could have been avoided with the use of adequate lubricants. More recently, estimates claim that the amount of energy wasted due to insufficient knowledge applied to the science of friction, lubrication, and wear resulted in roughly 0.4 % of the gross domestic product (GDP) being wasted [2]. In the United States, this means that over \$60.36 billion of the \$15.08 trillion GDP was wasted due to energy loss [3]. When considering the many macroscale applications that utilize lubricants such as internal combustion engines, turbines, hydraulic systems, compressors, vehicle and industrial gearboxes, and journal and thrust bearings as well as the various micro- and nanoscale applications and metal forming applications, it becomes easy to understand the importance that lubricants play in the compliance, effectiveness, and operation of many of these applications [4–6].

Within the lubrication market there are a vast number of applications which require specifically formulated lubricants that have given rise to the upwards of 10,000 different lubricants that satisfy more than 90 % of all lubricant applications worldwide [2]. Figure 1 analyzes the global lubrication market as of 2004, which consumed roughly 37.4 million tons of lubricant [2]. This figure illustrates how automotive and industrial lubricants are the most prevalent. Industrial lubricants amount to 32 % and were composed of 12 % hydraulic oils, 10 % other industrial oils, 5 % metalworking fluids, 3 % greases, and 2 % industrial gear oils [7, 8]. The

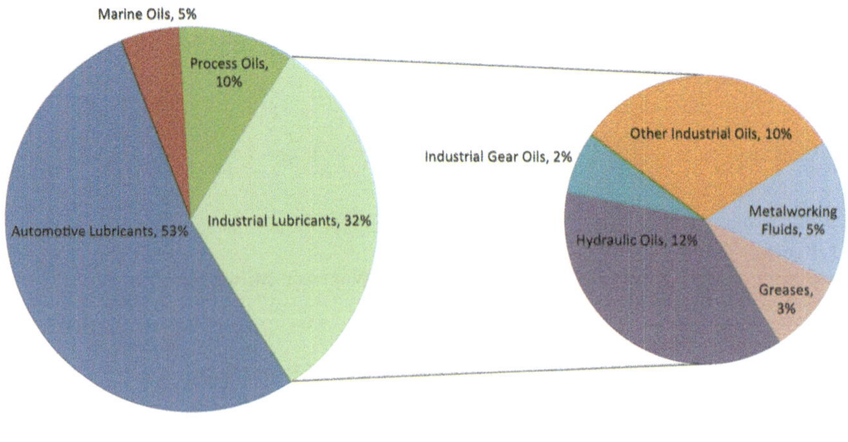

Fig. 1 2004 Worldwide lubrication consumption [2]

10 % of other industrial oils within the industrial lubricants section consist of a wide range of lubricants such as air and gas compressor oils, bearing and circulating system oils, refrigerator compressor oils, and steam and gas turbine oils. In the automotive lubricants section, the most commonly used liquid lubricants were engine oils (petrol and diesel engine oils), automatic transmission fluids, gearbox fluids, brake fluids, and hydraulic fluids.

2 History of Eco-friendly Lubricants

Lubricants originally consisted of natural oils and fats derived from plant- and animal-based raw materials which date back to 1400 B.C. The modern lubrication market developed after the first oil well was drilled in 1959 in Titusville, PA. Since then, lubricants have evolved from mineral oils to petrochemically modified synthetic oils arriving in the 1960s, to today's bio-based eco-friendly lubricants harvested from raw materials derived from the oleochemical industry. Recently, bio-based lubricants have begun to seek prominence for their environmental-friendliness and superior tribological properties. The current trend in the lubrication industry is to develop more bio-based lubricants due to estimates indicating that nearly 50 % of all lubricants sold worldwide pollute the environment, through spillage, evaporation, and total loss applications [9]. An example of lubrication pollution is that of the diesel engine particulate emissions, where approximately one-third of the engine oil vaporizes thus polluting the atmosphere. The large quantity of lubricant loss into the environment is the reason behind the development of eco-friendly lubricants. Although the lubrication market is shifting to become more environmentally responsible by reducing the use of petroleum-based lubricants due to concerns of protecting the environment, depletion of oil reserves, and increases in oil price, mineral oil remains to be the largest constituent and the foundation to most lubricants [9, 10].

The use of biolubricants derived from plant oils and animal fats dates back to antiquity. Scientists have known for centuries that biolubricants provide favorable friction and wear properties. Since the beginning of the twentieth century, investigations into the properties of bio-based oils have received significant attention due to the fact that 50 % of all lubricants worldwide end up in the environment through usage, spill, volatility, or improper disposal [11–13]. Of these lubricants entering the environment, 95 % are derived from petroleum-based oils and are detrimental to many biological ecosystems [14]. More still, within North America alone, over 100 million gallons of toxic lubricants drip, spill, and leak into the environment annually. With the advent of petroleum-based oils in the mid-1800s, the use of bio-based oils as lubricants began to decline dramatically. Recently, there has been a resurgence of eco-friendly lubricants due to increased environmental efforts to reduce the use of petroleum-based lubricants in addition to the depletion of oil reserves, increases in oil price, and rises in lubricant disposal costs [15, 16].

When compared to petroleum-based lubricants, bio-based lubricants have a higher lubricity, lower volatility, higher shear stability, higher viscosity index, higher load carrying capacity, and superior detergency and dispersancy [9, 17, 18], therefore they are excellent alternatives to petroleum-based oils. Traditional eco-friendly lubricants are typically derived from naturally occurring organic substances whose properties and utility vary based on biological factors such as nutrient availability, climate, light, temperature, and water [19–21]. Despite these favorable attributes, the largest drawbacks to many bio-based oils are their poor thermal-oxidative stability, high pour points, and inconsistent chemical composition, which have led to the development of chemically modified synthetic biolubricants, the use of stabilizing additives, and ionic liquids [22, 23].

2.1 Present: Eco-friendly Lubricants

The emphasis placed on bio-based lubricants is a result of the increase in demand for eco-friendly lubricants that are less toxic to the environment, renewable, and provide feasible and economical alternatives to traditional lubricants. Currently, the interest surrounding liquid lubricants derived from various bio-based feedstocks is focused on the use of eco-friendly lubricants derived from plant-based oils. This is the result of the chemical composition consisting of triacylglycerol molecules made up of esters derived from glycerol and long chains of polar fatty acids. It is these fatty acids that are desirable in boundary lubrication for their ability to adhere to metallic surfaces due to their polar carboxyl group, remain closely packed, and create a monolayer film that is effective at reducing friction and wear by minimizing the asperity contact [24, 25]. Much of the work with eco-friendly lubricants has concentrated on understanding the fundamentals of saturated and unsaturated fatty acids with the bulk of the attention focusing on the use of natural oils as neat lubricants, fatty acids as additives in mineral oils, and bio-based feedstock for chemically modified lubricants [12, 17, 26]. Recently, eco-friendly lubricants are finding uses as carrier fluids for lamellar powder additives in sliding contact [9, 12, 27, 28].

Eco-friendly lubricants composed of environmentally benign lamellar powders such as boric acid (H_3BO_3) and hexagonal boron nitride (hBN) are well-known solid lubricants for their low interlayer friction, ability to form protective boundary layers, and accommodate relative surface velocities [17, 28]. As with many lamellar powders, atoms on the same plane form layers through strong covalent bonds. These layers themselves are held together through the weak van der Waals force, providing the minimal shear resistance, and enabling the low interlayer friction [29]. Lamellar powders are effective in a broad range of environments of extreme pressure and temperature as well as various applications from automotive to aerospace to lower friction and minimize wear [9, 21]. An important property of boron-derived lamellar powders is that they are environmentally benign and inert to most chemicals making them attractive performance enhancing additives to bio-based oils. Experiments have shown that these lamellar particles can be forced out of the contact zone in sliding contact and therefore adding them to natural oils such as canola oil creates a superior eco-friendly lubricant [12, 30]. This new class of eco-friendly lubricant maintains the properties of the powder additives to coalesce and fill in the asperity valleys, thereby establishing a thin, smooth, solid lamellar film between the contacting surfaces, thus decreasing the friction coefficient, wear rate, and surface roughness [9, 31, 32]. In addition, these lubricants maintain boundary lubrication characteristics by establishing the fatty acid adsorption film that thwarts metal-to-metal contact.

2.2 Drawbacks to Eco-friendly Lubricants

The use of lubricants composed of natural plant oils or solid lubricants have their merits; however, they do have their limitations, which have stifled their ability to be widely accepted within the lubrication industry. The drawbacks to these lubricants are summarized below. For natural oils, they suffer from thermal-oxidative instability, high pour points, inconsistent chemical composition, hydrolytic instability, and a severe susceptibility to biological deterioration. For lamellar powders, they suffer from concentration optimization (which affects their price making these lubricants expensive), unwanted abrasive behavior due to particle size and shape, particles can settle out of the colloidal suspension rendering them useless, large particles can block tubes and capillaries within critical engine parts, and they can clog oil filters in circulatory lubrication systems. These shortcomings of traditional eco-friendly lubricants ultimately cause economic issues where the lubricants themselves can become very expensive when modifying their properties for many thousands of potential applications.

3 Future: Eco-friendly Lubricants

3.1 Ionic Lubricants

A new type of eco-friendly lubricant, ionic liquids, is beginning to gain attention. Ionic liquids (ILs) were originally a novel class of solvents typically consisting of an organic cation in combination with any of a wide variety of organic or inorganic anions, exhibit a number of unique and useful characteristics, including high thermal stability, low melting point, a broad liquidus range, and negligible vapor pressure. The last of these properties, particularly minimizing solvent losses due to volatilization (i.e., fugative emissions), have led many to regard ionic liquids as "green solvents," and over the last decades, they have been evaluated in a wide range of applications, including the fabrication of dye-sensitized solar cells [33], the preparation of electrolytes for electrochemical storage devices [34] and batteries, the electrodeposition of metals [35], the recovery of metal ions from aqueous solutions via liquid–liquid extraction (LLE) [36, 37], and in the development of separation processes for various organic compounds [38]. Many of the same properties that make ILs useful in these applications also make them good candidates as high-performance lubricants [39].

The use of ionic liquids as lubricants was first reported in 1961, when fluoride-containing molten salts (i.e., LiF and BeF_2) were subjected to high-temperature (650–815 °C) bearing tests [40]. Nearly four decades later, low melting analogs of classical molten salts, room temperature ionic liquids (RTILs), were first evaluated as synthetic lubricating fluids [41]. Since this time, considerable attention has been devoted to the utilization of ILs as lubricants. Three main applications have been most extensively explored: the use of ILs as base oils, as additives, and as thin films [42]. When employed as base oils, ILs have been reported to exhibit good tribological performance for steel/steel, steel/copper, steel/aluminum, ceramic/ceramic, and steel/ceramic sliding pairs [43–56]. The negligible vapor pressure of ILs makes them good candidates for use under vacuum and in spacecraft applications [42]. ILs are also effective as additives to the main lubricant (e.g., mineral oils), because of their tendency to form strong boundary films, that enhance the tribological performance of the base lubricant [42, 57, 58]. Thin-film lubrication employing ILs has been studied by many researchers with the goal of replacing conventional perfluoropolyether (PFPE) lubricants [54, 55, 59–63].

Although the chemical structure of the cationic and anionic substituents of an IL can vary greatly, the most commonly studied ILs in tribological processes have been those containing a tetrafluoroborate (BF_4^-) or hexafluorophosphate (PF_6^-) anion [64, 65], the result of the superior tribological properties that boron- and/or phosphorus-containing compounds often exhibit under the high pressures and elevated temperatures that lubricants can encounter [66–69]. The frequent use of boron- and phosphorus-containing ILs as lubricants does not imply that either of them is optimum. Rather, ILs based on these anions are commonly studied because they are readily available and low cost [70]. In fact, BF_4^- and PF_6^- have been found to cause corrosion of steel under humid conditions. Moreover, other hydrophobic anions, such as bis(trifluoromethanesulfonyl)amide (TFSA) and tris(tetrafluoroethyl)trifluorophosphate (FAP), actually exhibit better tribological properties for steel–steel contact [39, 70, 71]. In general, as the hydrophobicity of the anion increases, both the thermo-oxidative stability and the tribological properties improve [70].

Among the many possible IL cations, the imidazolium ion has probably been studied in the most detail, a result of the high thermal stability of imidazole-based rings [72]. Additionally, the chain length on the imidazolium cation can be readily altered. Increasing the chain length to make the IL more hydrophobic will decrease the friction coefficient in a manner similar to that observed when the anion is made more hydrophobic. In contrast to the improvement in thermo-oxidative stability observed with hydrophobic anions, however, a decrease in stability is observed with more hydrophobic cations [70]. Nonetheless, ILs with longer alkyl chains and lower polarity have been reported to have excellent tribological properties from low to high temperature (−30 to 200 °C) [73]. Other ILs have been studied with the goal of improving their tribological properties include phosphonium [74–76] and ammonium [53, 54, 77–79].

3.2 Room Temperature Ionic Liquid Lubricants

As the industrial marketplace continues to become more ecologically focused with much of the attention centered on novel approaches to achieve efficient energy conservation and sustainability, new classes of green lubricants are being developed that represent the future potential of eco-friendly lubricants. Much of the development aims at creating environmentally friendly lubricants that contain many of the properties of the aforementioned eco-friendly lubricants such as polar molecules (similar to the fatty acids), lamellar crystal structure (like the solid lubricants), derived from bio-based feedstock (natural plant-based oils), high thermal-oxidative stability, physicochemical consistency (which natural oils inherently lack), superior lubricity with minimal wear, and require minimal use of additives [80]. Ionic liquids, particularly those that are fluid at room temperature, represent a promising new class of eco-friendly lubricants that show potential to improve the limitations associated with current petroleum-based oils, bio-based oils, and solid powder additives [42, 43, 81, 82].

Room temperature ionic liquids consist of combinations of a bulky, asymmetric organic cation, and an appropriate organic anion with melting points below 100 °C and a liquid range beyond 300 °C [83, 84]. The atomic structure of an IL is shown in Fig. 2. This structure resembles a lamellar solid crystal structure, except with ILs the anions and the cations form ionic bonds to creating layers and these layers are held together with the weak van der Waals forces [85]. This structure provides ILs with their liquid lamellar crystal structure [86]. Aforementioned, ILs exhibit a number of unique and useful properties that make them well suited as the basis of a new family of lubricants and initial research has already begun investigating the properties of ILs [39, 43, 81, 87–89]. The appeal of ILs as lubricants becomes even more evident when one considers their many potential advantages over other lubricants including: (1) a broad liquid range (low melting and high boiling point); (2) negligible vapor pressure; (3) nonflammability and noncombustibility; (4) superior thermal stability; (5) high viscosity; (6) miscibility and solubility; (7) environmentally benign (nontoxic); (8) lamellar-like liquid crystal structure; (9) long polar anion-cation molecular chains; and (10) economical costs [35, 39, 42, 70, 90, 91].

Additionally, ionic liquids have a consistent and easily tailorable chemical composition that affords them the ability to provide the level of thermal-oxidative stability and lubricity required for a variety of applications in the aerospace, automotive, manufacturing, and magnetic storage industries [35, 44]. The consistent

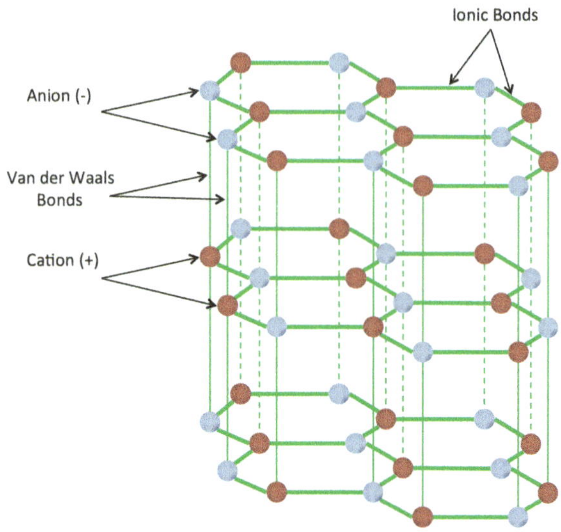

Fig. 2 Atomic strucutre of an ionic liquid

chemical composition allows ILs to have physicochemical properties that are readily reproducible. Furthermore, they can be designed to be eco-friendly by selecting both the cationic and anionic constituents to be nontoxic. In many instances, the ILs can be prepared from nonpetroleum resources. Lastly, their capacity to overcome the variety of environmental, cost, and performance challenges faced by conventional lubricants makes them a potentially attractive alternative eco-friendly lubricant [42, 83].

The possibility of preparing an ionic liquid capable of functioning as an efficient lubricant, while exhibiting a variety of other useful properties is a result of the physicochemical characteristics, inherent tunability, and structural diversity of these novel compounds. Regarding the latter point, it has been estimated that as many as 10^{18} different combinations of anion and cation moieties are possible [92, 93]. Clearly, this vast assortment of possibilities can pose a significant challenge in ionic liquid design. As the number of desired properties increases, the number of possible candidate ILs declines dramatically. Here, for example, the desire for an environmentally friendly bio-based lubricant means that the use of highly fluorinated anions is unacceptable [94]. Instead, the use of carboxylic anions based on common food additives (e.g., benzoate$^-$ and salicylate$^-$ are well-known preservatives) or artificial sweeteners (e.g., saccharinate$^-$) are utilized. Similar considerations guide the choice of the cation and suggest that trihexyl(tetradecyl)phosphonium salts (i.e., $P^+_{666,14}$), some of which have been found to exhibit antimicrobial and biodegradable properties, can satisfy many of the desired criteria [95]. Along these same lines, the objective of employing renewable feedstocks for the preparation of the ILs suggests the use of certain 1,3-dialkylimidazolium cations, such as can be derived from fructose or other bio-based feedstock [96–100].

4 Case Study: Room Temperature Ionic Liquids

To assess the potential of using ILs as base lubricants, two studies were investigated consisting of imidazolium and phosphonium ILs with carboxylate anions for their ability to address the performance challenges of traditional lubricants [58, 101]. In the first study, a series of experiments were conducted with two conventional ionic liquids, a phosphonium-based ($P_{666,14}Tf_2N$) and an imidazolium-based ($C_{10}mimTf_2N$) ionic liquid that were both mixed with avocado oil in five different proportions as shown in Table 1 to investigate their use as an additive versus a base fluid [101]. In the second study, another set of experiments were conducted by interchanging the cation-anion moieties of the phosphonium-based and imidazolium-based ionic liquids [58]. Multiple IL lubricants having different ion pairs were then compared with various vegetable oils and commercial lubricants. In both studies, pin-on-disk tests were conducted to characterize the tribological performance of ionic liquids as possible lubricants and in particular investigate the performance of bio-based environmentally benign ionic liquid lubricants.

4.1 Study 1: Ionic Liquids as Additives in Natural Oils

Figure 3 shows the variation of the COF for different mixtures of phosphonium-based ILs and avocado oil. In Fig. 3a it can be seen that the lubricant mixtures generally decrease with sliding distance and eventually reach a steady state value at a sliding distance of approximately 1000 m. Figure 3b shows the final COF values at the completion of the tests. It can be seen that as the graph moves from left to right the amount of the IL sequentially decreases from 100 to 0 % in 25 % decrements. This means that on the left, the 100 % IL is a lubricant composed entirely of $P_{666,14}Tf_2N$ and on the right, the 0 % IL is a lubricant composed entirely of avocado oil. The lubricant mixtures in the middle are composed of combinations of the two base fluids. Table 1 displays the friction results at the completion of the tests. The correlation between the friction results and the composition of the IL in

Table 1 Ionic liquid and avocado oil lubricant mixture—tribological results at the completion of the tests

Mixture number	Ionic liquid percentage (%)	Natural oil percentage	Notation	$P_{666,14}Tf_2N$, COF	$C_{10}mimTf_2N$, COF
(1)	0	0.0168	0 % IL	0.0155	0.0135
(2)	25	0.0259	25 % IL	0.0223	0.0185
(3)	5	0.0355	50 % IL	0.0381	0.0335
(4)	75	0.0417	75 % IL	0.0442	0.0401
(5)	100	0.0512	100 % IL	0.0498	0.0507
Correlation coefficient, *R*-value				−0.982	−0.998

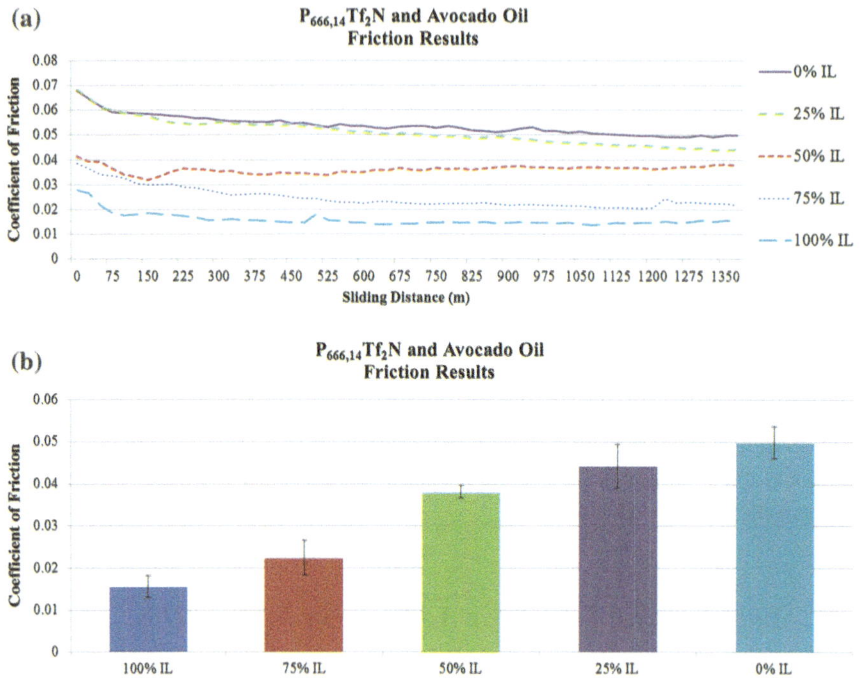

Fig. 3 Variation of the coefficient of friction for different mixtures of phosphonium-based ionic liquid and avocado oil. **a** Lubricant mixtures with sliding distance and **b** lubricant mixtures at the completion of tests

the base fluid is strong with a negative correlation coefficient (*R*-value) of −0.982 as shown in Table 1. This indicates that as the percentage of the IL increases in the lubricant mixture the COF decreases.

Having similar results to the phosphonium-based IL and avocado oil lubricant mixtures, Fig. 4 shows the variation of the COF for different mixtures of imidazolium-based IL and avocado oil lubricant mixtures. In Fig. 4a it can be seen that the lubricant mixtures continue to decrease with sliding distance and eventually reach a steady state value at a sliding distance about of 1000 m. Figure 4b shows the final COF values at the completion of the tests. Similarly, it can be seen that as graph moves from left to right the amount of the IL sequentially decreases from 100 to 0 %. On the left the 100 % IL is a lubricant composed entirely of $C_{10}mimTf_2N$ and on the right, the 0 % IL is a lubricant composed entirely of avocado oil. The lubricant mixtures in the middle are composed of combinations of the two base fluids. The correlation between the friction results and the composition of the IL in the base fluid is strong with a negative correlation coefficient (*R*-value) of −0.998, as shown in Table 1. This indicates that as the percentage of the IL increases in the lubricant mixtures the COF decreases.

The friction results indicate that the presence of either the phosphonium-based or imidizolium-based IL as an additive improves the COF. Similar results were

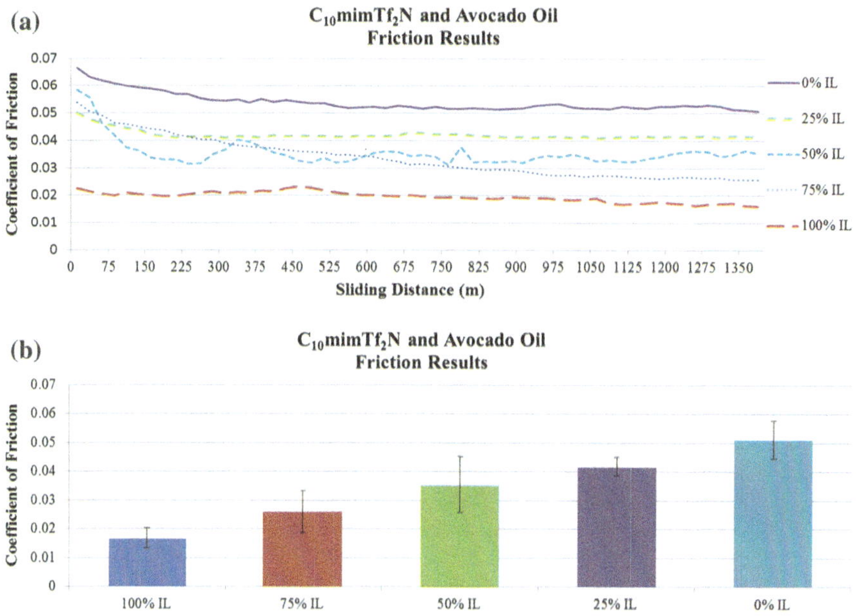

Fig. 4 Variation of the coefficient of friction for different mixtures of imidazolium-based ionic liquid and avocado oil. **a** Lubricant mixtures with sliding distance and **b** lubricant mixtures at the completion of tests

witnessed with wear rates. As detailed in Table 1, as the amount of the IL additive increases to the point where it becomes the majority fluid, effectively becoming the base fluid, it continues to impart superior friction properties that outperform the avocado oil. This study demonstrates that ionic liquids can not only be considered as additives in lubricants, but they can also be considered as base fluids in the form of neat lubricants. In the following investigation, various ILs having different anion-cation moieties will be examined for their friction properties. The IL lubricants will be compared to other natural oils and well-known commercial lubricants from both petroleum-based and bio-based feedstocks.

4.2 *Study 2: Ionic Liquid Anion-Cation Moiety Manipulation*

Figure 5 shows the variation of the coefficient of friction at the completion of the tests for different anion-cation moieties separated into two groups. The first group shown in Fig. 5a, shows a cation study of different ionic liquids all with the same Tf_2N anion. Here, an investigation into the influence of cation chain length was performed with the five imidazolium cations having chain lengths of 10, 8, 6, 5, and

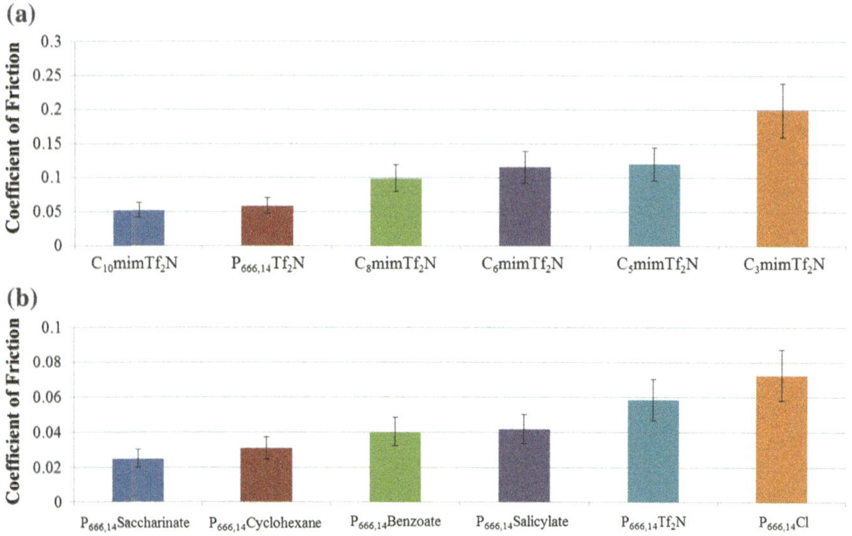

Fig. 5 Variation of the coefficient of friction at the completion of the tests for different anion-cation moieties. **a** $X^{+}Tf_2N$ cation study and **b** $P_{666,14}X^{-}$ anion study

3 carbons as well as the phosphonium cation, $P_{666,14}$ having a chain length of 14 carbon atoms. It can be seen in Fig. 5a and similarly in Table 2 that as the COF decreases, the cation chain length increases, with a strong negative correlation coefficient, *R*-value of −0.878. It can be seen that having 10 or 14 carbon atoms produces similar friction results with a 10 % difference maintaining values of 0.0526 and 0.0586, respectively. The COF percent difference between 10 and 14 carbon atoms is approximately 10 %, whereas the percent difference between 10 and 8 carbon atoms for two imidazolium-based cations is 47 %. Furthermore, the COF percent difference between 8 and 3 carbon atoms is 50 %. The relative percent differences indicate the importance of larger alkyl chains in the cation help to promote a thicker adsorbed monolayer film that is more effective at minimizing asperity contact. These results are in agreement with previous investigations of ionic liquids [42].

Table 2 $X^{-}Tf_2N$ cation chain length investigation results

Ionic Liquid	Cation	Anion	Chain length	COF
$C_{10}mimTf_2N$	$C_{10}mim$	Tf_2N	10	0.0526
$P_{666,14}Tf_2N$	$P_{666,14}$	Tf_2N	14	0.0586
C_8mimTf_2N	C_8mim	Tf_2N	8	0.099
C_6mimTf_2N	C_6mim	Tf_2N	6	0.116
C_5mimTf_2N	C_5mim	Tf_2N	5	0.120
C_3mimTf_2N	C_3mim	Tf_2N	3	0.199
Correlation coefficient, *R*-value				−0.878

Table 3 $P_{666,14}X^+$ anion ring size investigation results

Ionic Liquid	Cation	Anion	Ring size	COF
$P_{666,14}$Saccharinate	$P_{666,14}$	Saccharinate	2	0.025
$P_{666,14}$Cyclohexane	$P_{666,14}$	Cyclohexane	1	0.0309
$P_{666,14}$Benzoate	$P_{666,14}$	Benzoate	1	0.0403
$P_{666,14}$Salicylate	$P_{666,14}$	Salicylate	1	0.0419
$P_{666,14}$Tf2N	$P_{666,14}$	Tf2N	0	0.0586
$P_{666,14}$Cl	$P_{666,14}$	Cl	0	0.0727
Correlation coefficient, *R*-value				−0.917

The second group shown in Fig. 5b reveals the anion study of different ionic liquids, all with the same $P_{666,14}$ cation. This study examines the influence of anion ring size on the phosphonium-based ionic liquids. The anions invested are shown in Table 3 along with their corresponding ring size and final COF values. It can be seen in Fig. 5b and in Table 3, that as the COF decreases, the anion ring size increases, with a strong negative correlation coefficient, of −0.917. It can be seen that the saccharinate with two aromatic rings has the lowest COF value followed by cyclohexane carboxylate. The cyclohexane has a 6-vertexed ring that does not conform to the shape of a perfect hexagon, thus its nonplanar shape is often considered a 3D chair or boat conformation. The benzoate and the salicylate anions demonstrate similar properties as they have very similar aromatic molecular structures consisting of only one ring and a difference of one hydroxyl group as observed in Table 3. The Tf_2N and Cl anions have no ring shape and are included for comparative purposes. Interestingly, the ring-shaped anions (saccharinate, cyclohexane, benzoate, and salicylate) maintain a significantly lower COF value than the C_{10}mim cation. The difference between the $P_{666,14}$Salicylate and the C_{10}mimTf_2N reveals a 20 % difference in the friction values and this difference can only increase as the ring size increases. Therefore, the influence of larger ring sized anions tends to improve upon the effect that longer alkyl chain length cations have on lowering friction. Similar results were observed with wear rate as well where studies of the pin worn surfaces reaffirm that aromatic anions and long alkyl chains length cations are important for improved tribological properties. It can be inferred that this is due to density of the monolayer to remain tightly packed on a surface, thus covering more of the surface to prevent unwanted asperity contact and aiding to minimize wear.

4.3 Environmentally Friendly Ionic Liquid Lubricants

The anion-cation investigations thus far have shown that different anionic and cationic constituents influence the performance of ionic liquids. Within the combinations of anion-cation moieties investigated, those ILs with superior tribological properties can serveve as practical lubricants. Further examination will compare the

Table 4 Classification of investigated lubricants based on source

Eco-friendly ILs	Conventional ILs	Bio-based oils	Petroleum-based oils
$P_{666,14}$Benzoate	$P_{666,14}$Cyclohexane	Peanut	Synthetic Motor oil
$P_{666,14}$Saccharinate	$P_{666,14}Tf_2N$	Avocado	–
$P_{666,14}$Salicylate	$P_{666,14}Cl$	Canola (rapeseed)	–
	$C_{10}mimTf_2N$	Vegetable (soybean)	–
	C_8mimTf_2N	Commercial biolubricant	–
	C_6mimTf_2N	–	–
	C_5mimTf_2N	–	–
	C_3mimTf_2N	–	–

tribological properties of the ILs in particular, the eco-friendly ionic liquids to conventional ILs, bio-based (natural) oils, and petroleum-based oils. Table 4 classifies the lubricants into the four categories based on their source and Table 5 compiles all the friction values for the lubricants tested.

Figure 6 depicts the final COF values at the completion of the tests for all of the lubricants studied in this investigation. The ionic liquid lubricants maintain superior tribological properties when compared to all other lubricants tested. More specifically, the eco-friendly ILs ($P_{666,14}$Saccharinate, $P_{666,14}$Salicyate, and $P_{666,14}$Benzoate) exhibit lower coefficient of friction values better than all conventional ILs (except $P_{666,14}$Cyclohexane), bio-based oils, and petroleum-based oil. Two trends emerge when analyzing the lubricant chemical composition and the friction values. As denoted previously in Fig. 5, the COF decreases as the alkyl

Table 5 Classification of investigated lubricants

Lubricant	Classification	COF
$P_{666,14}$Saccharinate	Eco-friendly IL	0.025
$P_{666,14}$Cyclohexane	Conventional IL	0.031
$P_{666,14}$Benzoate	Eco-friendly IL	0.040
$P_{666,14}$Salicylate	Eco-friendly IL	0.042
$C_{10}mimTf_2N$	Conventional IL	0.053
$P_{666,14}Tf_2N$	Conventional IL	0.059
Vegetable (soybean) oil	Bio-based oils	0.069
Peanut oil	Bio-based oils	0.072
$P_{666,14}Cl$	Conventional IL	0.073
Avocado oil	Bio-based oils	0.073
Canola (rapeseed) oil	Bio-based oils	0.096
C_8mimTf_2N	Conventional IL	0.099
Commercial biolubricant	Bio-based oils	0.106
C_6mimTf_2N	Conventional IL	0.116
C_5mimTf_2N	Conventional IL	0.120
Synthetic oil	Petroleum-based oil	0.173
C_3mimTf_2N	Conventional IL	0.199

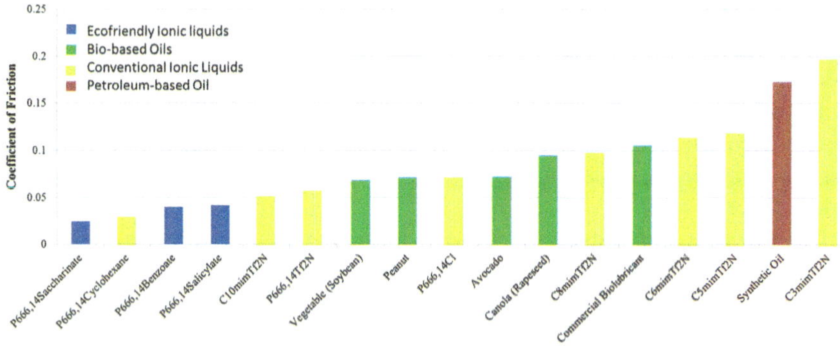

Fig. 6 Variation of the coefficient of friction for different lubricant mixtures of eco-friendly ionic liquids, natural oils, conventional ionic liquids, and commercial lubricants at the completion of tests

chain length of the cation increases and as the aromatic ring size of the anion increases. These trends remain true with the eco-friendly ILs, as they consist of the largest cation with 14 carbon atoms in the alkyl chain and the largest anions with single or double aromatic rings that enable them to have low friction. Comparing the eco-friendly ILs to the bio-based natural oils, the eco-friendly ILs outperform the bio-based oils in all circumstances. This is speculated to be caused by the resilient monolayers composed of both anion and cation molecules. Here, the ionic liquids are able to maintain lower COF values due to their polar molecules that can adhere to the charged metallic surfaces, establish a monolayer, remain tightly packed, and minimize asperity contact as well as benefit from their lamellar liquid crystal structure that affords the ILs the reduced internal resistance during shearing in the tribo-interface. On the contrary, the bio-based oils maintain high lubricity due to their fatty acid composition consisting of over 70 % of oleic acid (C18:1) and linoleic acid (C18:2) [25]. These fatty acids having 18 carbon atoms and one or two double bonds can form a monolayer that promotes low friction, however, their properties are susceptible to rapid oxidation and chemical inconsistency, which can impede their tribological performance. The synthetic motor oil tested exhibits the highest friction. The friction results are important because they directly reveal the advantages of eco-friendly ionic liquids that make them well suited as a new class of greener lubricants.

5 Conclusions

This chapter explores the history of eco-friendly lubricants from their past as natural impeller pressed plant-based oils, to more complex natural oils with environmentally benign solid particle additives, and now shedding light on a promising new class of environmentally friendly lubricants known as room temperature ionic liquid lubricants. It has been shown that their lamellar-like liquid crystal structure improves lubricity in comparison to traditional eco-friendly lubricants. The dipolar structure allows cations and anions to adsorb on charged worn metal surfaces facilitating self-assembling monolayers that create boundary films that minimize wear, reduce friction, and improve component operation. Eco-friendly ILs were presented and demonstrated to be feasible in design and provide superior tribological properties.

The lubrication industry continues to make new strides toward sustainable eco-friendly lubricants with properties that will lower friction and wear, thereby improving system efficiency and ultimately conserving energy. Continuing this trend, future lubricants formulated from bio-based feedstock should offer the following advantages over petroleum-based oils: a higher lubricity lending to lower friction losses and improved efficiency, affording more power output and better economy. As oil prices rise, environmental awareness grows, and the demand for renewable and sustainable lubricants increases, eco-friendly lubricants will begin to seek prominence. For this reason, fundamental research is an important step to the macroscale development, economical competence, and industrial use of biolubricants for energy conservation and sustainability. Room temperature ionic liquid lubricants represent a new class of novel "greener" lubricants that are nontoxic, obtainable from sustainable (nonpetroleum) resources, and environmentally friendly. They have the potential to satisfy the combination of environmental, health, economic, and performance demands of modern lubricants. Their ability to be tunable establishes them as 'designer' lubricants, where the optimization of cation-anion moiety facilitates energy conservation through superior tribological performance.

References

1. Bannister KE (1996) Lubrication for industry. Industrial Press, New York
2. Mang T, Dresel W (2006) Lubricants and lubrication. Wiley, Weinheim
3. United States (1997) Central intelligence A. Central Intelligence Agency, The CIA world fact book. Washington DC
4. Reeves C, Menezes PL, Lovell M, Jen T-C (2013) Macroscale applications in tribology. In: Menezes PL, Nosonovsky M, Ingole SP, Kailas SV, Lovell MR (eds) Tribology for scientists and engineers. Springer, New York, pp 881–919
5. Reeves C, Menezes PL, Lovell M, Jen T-C (2013) Microscale applications in tribology. In: Menezes PL, Nosonovsky M, Ingole SP, Kailas SV, Lovell MR (eds) Tribology for scientists and engineers. Springer, New York, pp 921–948
6. Menezes PL, Reeves C, Kailas S, Lovell M (2013) Tribology in metal forming. In: Menezes PL, Nosonovsky M, Ingole SP, Kailas SV, Lovell MR (eds) Tribology for scientists and engineers. Springer, New York, pp 783–818
7. Bartz WJ (2006) Automotive and industrial lubrication 15th International Colloquium Tribology. TAE, Ostfildern
8. Lingg G, Gosalia A (2008) The dynamics of the global lubricants industry: markets, competitors and trends. In: Technische Akademie Esslingen international tribology colloquium proceedings, p 16

9. Reeves CJ, Menezes PL, Jen T-C, Lovell MR (2012) Evaluating the tribological performance of green liquid lubricants and powder additive based green liquid lubricants. STLE annual meeting and exhibition. STLE, St. Louis
10. Menezes PL, Reeves C, Lovell M (2013) Fundamentals of Lubrication. In: Menezes PL, Nosonovsky M, Ingole SP, Kailas SV, Lovell MR (eds) Tribology for scientists and engineers. Springer, New York, pp 295–340
11. Lundgren SM, Persson K, Mueller G, Kronberg B, Clarke J, Chtaib M et al (2007) Unsaturated fatty acids in alkane solution: adsorption to steel surfaces. Langmuir ACS J Surf Colloids 23:10598–10602
12. Lovell MR, Menezes PL, Kabir MA, Higgs CF III (2010) Influence of boric acid additive size on green lubricant performance. Philos Trans R Soc A Math Phys Eng Sci 368: 4851–4868
13. Lundgren SM, Ruths M, Danerlov K, Persson K (2008) Effects of unsaturation on film structure and friction of fatty acids in a model base oil. J Colloid Interf Sci 326:530–536
14. Schneider MP (2006) Plant-oil-based lubricants and hydraulic fluids. J Sci Food Agric 86:1769–1780
15. Deffeyes KS (2009) Hubbert's peak. Princeton (N.J.). Princeton University Press, Oxford
16. Goodstein DL (2004) Out of gas: the end of the age of oil, 1st edn. W.W. Norton, New York
17. Menezes PL, Lovell MR, Kabir MA, Higgs III CF, Rohatgi PK (2012) Green lubricants: role of additive size. In: Nosonovsky M, Bhushan B (eds) Green tribology. Springer, Berlin, pp 265–286
18. Bennion M, Scheule B (2010) Introductory foods. Prentice Hall, Upper Saddle River
19. Duzcukoglu H, Sahin O (2011) Investigation of wear performance of canola oil containing boric acid under boundary friction condition. Tribol Trans 54:57–61
20. Erdemir A (1990) Tribological properties of boric acid and boric acid forming surfaces: part 1, crystal chemistry and self-lubricating mechanism of boric acid. In: Society of tribologists lubrication engineers annual conference. Argonne National Labs, Denver
21. Lovell M, Higgs CF, Deshmukh P, Mobley A (2006) Increasing formability in sheet metal stamping operations using environmentally friendly lubricants. J Mater Process Technol 177:87
22. Kumar A, Sharma S (2008) An evaluation of multipurpose oil seed crop for industrial uses (Jatropha curcas L.): a review. Ind Crops Prod 28:1–10
23. Reeves CJ, Menezes PL, Lovell MR, Jen T-C (2015) Science and technology of environmentally friendly lubricants. Environmentally friendly and biobased lubricants. CRC Press, Boca Raton
24. Grushcow J (2005) High oleic plant oils with hydroxy fatty acids for emission reduction. In: World tribology congress III. American Society of Mechanical Engineers, Washington, DC, pp 485–486
25. Reeves CJ, Menezes PL, Jen T-C, Lovell MR (2015) The influence of fatty acids on tribological and thermal properties of natural oils as sustainable biolubricants. Tribol Int 90:123–134
26. Fox NJ, Tyrer B, Stachowiak GW (2004) Boundary lubrication performance of free fatty acids in sunflower oil. Tribol Lett 16:275–281
27. Grushcow J, Smith MA (2005) Next generation feedstocks from new frontiers in oilseed engineering. ASME Conf Proc 2005:487–488
28. Reeves CJ, Menezes PL, Lovell MR, Jen T-C (2013) The size effect of boron nitride particles on the tribological performance of biolubricants for energy conservation and sustainability. Tribol Lett 51:437–452
29. Reeves C, Menezes PL, Lovell M, Jen T-C (2013) Tribology of solid lubricants. In: Menezes PL, Nosonovsky M, Ingole SP, Kailas SV, Lovell MR (eds) Tribology for scientists and engineers. Springer, New York, pp 447–494
30. Liang H, Jahanmir S (1995) Boric acid as an additive for core-drilling of alumina. J Tribol Journal of Tribology. 1995;117

31. Reeves CJ, Jen TC, Menezes PL, Lovell MR (2015) The influence of surface roughness and particulate size on the tribological performance of bio-based multi-functional hybrid lubricants. Tribol Int Tribol Int 88:40–55
32. Reeves CJ, Menezes PL, Jen T-C, Lovell MR (2013) The effect of surface roughness on the tribological performance of environmentally friendly bio-based lubricants with varying particle size. In: STLE annual meeting and exhibition. STLE, Detroit
33. Li Y, Pang A, Wang C, Wei M (2011) Metal organic frameworks: promising materials for improving the open circuit voltage of dye-sensitized solar cells. J Mater Chem 21: 17259–17264
34. Passerini S, Alessandrini F, Appetecchi GB, Conte M (2006) Ionic liquid based electrolytes for high energy electrochemical storage devices. ECS Trans 1:67–71
35. Liu W, Ye C, Gong Q, Wang H, Wang P (2002) Tribological performance of room-temperature ionic liquids as lubricant. Tribol Lett 13:81–85
36. Garvey SL, Hawkins CA, Dietz ML (2012) Effect of aqueous phase anion on the mode of facilitated ion transfer into room-temperature ionic liquids. Talanta 95:25–30
37. Hawkins CA, Garvey SL, Dietz ML (2012) Structural variations in room-temperature ionic liquids: influence on metal ion partitioning modes and extraction selectivity. SEPPUR Sep Purif Technol 89:31–38
38. Zaijun L, Jie C, Haixia S, Jiaomai P (2007) Advance of room temperature ionic liquid as solvent for extraction and separation. Rev Anal Chem 26:109–153
39. Liu X, Zhou F, Liang Y, Liu W (2006) Tribological performance of phosphonium based ionic liquids for an aluminum-on-steel system and opinions on lubrication mechanism. Wear 261:1174–1179
40. Smith PG (1961) High-temperature molten-salt lubricated hydrodynamic. J Bear ASLE Trans 4:263–274
41. Ye C, Liu W, Chen Y, Yu L (2001) Room-temperature ionic liquids: a novel versatile lubricant. Chem Commun 2001:2244–2245
42. Zhou F, Liang Y, Liu W (2009) Ionic liquid lubricants: designed chemistry for engineering applications. Chem Soc Rev 38:2590–2599
43. Yao M, Liang Y, Xia Y, Zhou F (2009) Bisimidazolium ionic liquids as the high-performance antiwear additives in poly(ethylene glycol) for steel-steel contacts. ACS Appl Mater Interf ACS Appl Mater Interf 1:467–471
44. Phillips BS, Zabinski JS (2004) Ionic liquid lubrication effects on ceramics in a water environment. Tribol Lett 17:533–541
45. Wang H, Lu Q, Ye C, Liu W, Cui Z (2004) Friction and wear behaviors of ionic liquid of alkylimidazolium hexafluorophosphates as lubricants for steel/steel contact. Wear 256:44–48
46. Liu W, Ye C, Chen Y, Ou Z, Sun DC (2002) Tribological behavior of sialon ceramics sliding against steel lubricated by fluorine-containing oils. Tribol Int 35:503–509
47. Mu Z, Liu W, Zhang S, Zhou F (2004) Functional room-temperature ionic liquids as lubricants for an aluminum-on-steel system. Chem Lett 33:524–525
48. Lu Q, Wang H, Ye C, Liu W, Xue Q (2004) Room temperature ionic liquid 1-ethyl-3-hexylimidazolium-bis(trifluoromethylsulfonyl)-imide as lubricant for steelsteel contact. Tribol Int 37:547–552
49. Reich RA, Stewart PA, Bohaychick J, Urbanski JA (2003) Base oil properties of ionic liquids. Lubr Eng 59:16–21
50. Mu Z, Zhou F, Zhang S, Liang Y, Liu W (2005) Effect of the functional groups in ionic liquid molecules on the friction and wear behavior of aluminum alloy in lubricated aluminum-on-steel contact. Tribol Int 38:725–731
51. Jimenez AE, Bermudez MD, Iglesias P, Carrion FJ, Martinez-Nicolas G (2006) 1-N-alkyl - 3-methylimidazolium ionic liquids as neat lubricants and lubricant additives in steel-aluminium contacts. Wear 260:766–782
52. Jimenez AE, Bermudez MD, Carrion FJ, Martinez-Nicolas G (2006) Room temperature ionic liquids as lubricant additives in steel-aluminium contacts: Influence of sliding velocity, normal load and temperature. Wear 261:347–359

53. Omotowa BA, Phillips BS, Zabinski JS, Shreeve JM (2004) Phosphazene-based ionic liquids: synthesis, temperature-dependent viscosity, and effect as additives in water lubrication of silicon nitride ceramics. Inorg Chem 43:5466–5471
54. Yu G, Zhou F, Liu W, Liang Y, Yan S (2006) Preparation of functional ionic liquids and tribological investigation of their ultra-thin films. Wear 260:1076–1080
55. Yu B, Zhou F, Mu Z, Liang Y, Liu W (2006) Tribological properties of ultra-thin ionic liquid films on single-crystal silicon wafers with functionalized surfaces. Tribol Int 39:879–887
56. Xia Y, Wang S, Zhou F, Wang H, Lin Y, Xu T (2006) Tribological properties of plasma nitrided stainless steel against SAE52100 steel under ionic liquid lubrication condition. Tribol Int 39:635–640
57. Qu J, Bansal DG, Yu B, Howe JY, Luo H, Dai S et al (2012) Antiwear performance and mechanism of an oil-miscible ionic liquid as a lubricant additive. ACS Appl Mater Interf 4:997–1002
58. Reeves CJ, Menezes PL, Garvey SL, Jen TC, Dietz ML, Lovell MR (2013) The effect of anion-cation moiety manipulation to characterize the tribological performance of environmentally benign room temperature ionic liquid lubricants. In: STLE annual meeting and exhibition (STLE2013). Society of Tribologists and Lubrication Engineers, Detroit
59. Mo Y, Zhao W, Zhu M, Bai M (2008) Nano/Microtribological properties of ultrathin functionalized imidazolium wear-resistant ionic liquid films on single crystal silicon. Tribol Lett 32:143–151
60. Zhu M, Yan J, Mo Y, Bai M (2008) Effect of the anion on the tribological properties of ionic liquid nano-films on surface-modified silicon wafers. Tribol Lett 29:177–183
61. Palacio M, Bhushan B (2008) Ultrathin wear-resistant ionic liquid films for novel MEMS/NEMS applications. Adv Mater 20:1194–1198
62. Bhushan B, Palacio M, Kinzig B (2008) AFM-based nanotribological and electrical characterization of ultrathin wear-resistant ionic liquid films. J Colloid Interf Sci 317: 275–287
63. Palacio M, Bhushan B (2008) Nanotribological and nanomechanical properties of lubricated PZT thin films for ferroelectric data storage applications. J Vac Sci Technol A Vac Surf Films 26:768–776
64. Minami I, Inada T, Sasaki R, Nanao H (2010) Tribo-chemistry of phosphonium-derived ionic liquids. Tribol Lett 40:225–235
65. Zeng Z, Shreeve JM, Phillips BS, Xiao JC (2008) Polyfluoroalkyl, polyethylene glycol, 1,4-bismethylenebenzene, or 1,4-bismethylene-2,3,5,6-tetrafluorobenzene bridged functionalized dicationic ionic liquids: synthesis and properties as high temperature lubricants. Chem Mater 20:2719–2726
66. Shah F, Glavatskih S, Antzutkin ON (2009) Synthesis, physicochemical, and tribological characterization of S-Di-n-octoxyboron-O, O'-di-n-octyldithiophosphate. ACS Appl Mater Interf 1:2835–2842
67. Mosey NJ (2005) Molecular mechanisms for the functionality of lubricant additives. Science 307:1612–1615
68. Mangolini F, Rossi A, Spencer ND (2011) Chemical reactivity of triphenyl phosphorothionate (TPPT) with iron: an ATR/FT-IR and XPS investigation. J Phys Chem C 115:1339–1354
69. Shah F, Glavatskih S, Höglund E, Lindberg M, Antzutkin ON (2011) Interfacial antiwear and physicochemical properties of alkylborate-dithiophosphates. ACS Appl Mater Interf 3: 956–968
70. Minami I (2009) Ionic liquids in tribology. Molecules (Basel, Switzerland 14:2286–2305
71. Itoh T, Ishioka A, Hayase S, Kawatsura M, Watanabe N, Inada K et al (2009) Design of alkyl sulfate ionic liquids for lubricants. Chem Lett 38:64–65
72. Ohtani H, Ishimura S, Kumai M (2008) Thermal decomposition behaviors of imidazolium-type ionic liquids studied by pyrolysis-gas chromatography. Anal Sci Int J Jpn Soc Anal Chem 24:1335–1340

73. Jimâenez A-E, Bermâudez M-D (2007) Ionic liquids as lubricants for steel-aluminum contacts at low and elevated temperatures. Tribol Lett 26:53–60
74. Shah FU, Glavatskih S, MacFarlane DR, Somers A, Forsyth M, Antzutkin ON (2011) Novel halogen-free chelated orthoborate-phosphonium ionic liquids: synthesis and tribophysical properties. Chem Phys (Incorporating Faraday Transactions) 13:12865–12873
75. Sun J, Howlett PC, MacFarlane DR, Lin J, Forsyth M (2008) Synthesis and physical property characterisation of phosphonium ionic liquids based on P(O)2(OR)2 and P(O)2(R)2 anions with potential application for corrosion mitigation of magnesium alloys. Electrochim Acta 54:254–260
76. Weng L, Liu X, Liang Y, Xue Q (2007) Effect of tetraalkylphosphonium based ionic liquids as lubricants on the tribological performance of a steel-on-steel system. Tribol Lett 26:11–17
77. Minami I, Kamimura H, Mori S (2007) Thermo-oxidative stability of ionic liquids as lubricating fluids. J Synth Lubr 24:135–147
78. Kamimura H, Kubo T, Minami I, Mori S (2007) Effect and mechanism of additives for ionic liquids as new lubricants. Tribol Int 40:620–625
79. Zhao W, Mo Y, Pu J, Bai M (2009) Effect of cation on micro/nano-tribological properties of ultra-thin ionic liquid films. Tribol Int 42:828–835
80. Reeves CJ, Menezes PL, Lovell MR, Jen TC, Garvey SL, Dietz ML (2013) The tribological performance of bio-based room temperature ionic liquid lubricants: a possible next step in biolubricant technology. World tribology congress—5th. Society of Tribologists and Lubrication Engineers, Torino
81. Reeves CJ, Garvey SL, Menezes PL, Dietz ML, Jen TC, Lovell MR (2012) Tribological performance of environmentally friendly ionic liquid lubricants. In: ASME/STLE 2012 international joint tribology conference. STLE, Denver
82. Reeves CJ, Menezes PL, Lovell MR, Jen TC (2014) The effect of particulate additives on the tribological performance of bio-based and ionic liquid-based lubricants for energy conservation and sustainability. In: STLE (ed) STLE annual meeting and exhibition. STLE, Buena Vista
83. Freemantle M (2010) An Introduction to ionic liquids. RSC Pub, Cambridge
84. Matlack A (2010) Introduction to green chemistry. Taylor & Francis Group
85. Manahan SE (1994) Environmental chemistry. Lewis, Boca Raton
86. Suisse J-M, Bellemin-Laponnaz S, Douce L, Maisse-François AWR (2005) A new liquid crystal compound based on an ionic imidazolium salt. Tetrahedron Lett 46:4303–4305
87. Yao Y, Wang X, Guo J, Yang X, Xu B (2008) Tribological property of onion-like fullerenes as lubricant additive. Mater Lett 62:2524–2527
88. Xia Y, Sasaki S, Murakami T, Nakano M, Shi L, Wang H (2007) Ionic liquid lubrication of electrodeposited nickel-Si3N4 composite coatings. Wear 262:765
89. Battez HA, Alonso DB, Rodriguez RG, Viesca Rodriguez JL, Fernandez-Gonzalez A, Garrido AH (2011) Lubrication of DLC and tin coatings with two ionic liquids used as neat lubricant and oil additives. In: Proceedings of the STLE/ASME international joint tribology conference. American Society of Mechanical Engineers, Los Angeles
90. Bermúdez MD, Jiménez AE, Sanes J, Carrión FJ (2009) Ionic liquids as advanced lubricant fluids. Molecules 14:2888–2908
91. Xue H, Tong ZF, Wei FY, Qing SG (2008) Crystal structure of room-temperature ionic liquid 1-butyl-isoquinolinium gallium tetrachloride [(BIQL)GaCl4]. Chem Rec 11:90–94
92. Canter N (2005) Evaluating ionic liquids as potential lubricants. Tribol Lubr Technol 61:15–17
93. Sheldon RA, Arends I, Hanefeld U (2007) Green chemistry and catalysis. Wiley, Weinheim
94. Wang H, Malhotra SV, Francis AJ (2011) Toxicity of various anions associated with methoxyethyl methyl imidazolium-based ionic liquids on Clostridium sp. Chemosphere 82:1597–1603
95. Atefi F, Garcia MT, Singer RD, Scammells PJ (2009) Phosphonium ionic liquids: design, synthesis and evaluation of biodegradability. Green Chem 11:1595–1604

96. Handy ST (2003) Greener solvents: room temperature ionic liquids from biorenewable sources. Eur J Chem 9:2938–2944
97. Gathergood N, Scammells PJ, Garcia TM (2006) Biodegradable ionic liquids part III. The first readily biodegradable ionic liquids. Green Chem 8:156–160
98. Gathergood N, Garcia TM, Scammells PJ (2004) Biodegradable ionic liquids: part I. Concept, preliminary targets and evaluation. Green Chem 6:166–175
99. Corma A, Iborra S, Velty A (2007) Chemical routes for the transformation of biomass into chemicals. Chem Inform 38
100. Zhang ZC (2013) Catalytic transformation of carbohydrates and lignin in ionic liquids. WENE Wiley interdisciplinary reviews: energy and environment
101. Reeves CJ, Jen T-C, Garvey SL, Dietz ML, Menezes PL, Lovell MR (2014) The effect of phosphonium-and imidazolium-based ionic liquids as additives in natural oil: an investigation of tribological performance

Chapter 3
New Emerging Self-lubricating Metal Matrix Composites for Tribological Applications

Emad Omrani, Afsaneh Dorri Moghadam, Pradeep L. Menezes and Pradeep K. Rohatgi

Abstract Self-lubricating metal matrix composites (SLMMCs) are an important category of engineering materials that are increasingly replacing a number of conventional materials in the automotive, aerospace, and marine industries due to superior tribological properties. Implementing self-lubricating composites into different operating systems is a solution to reduce the use of external toxic petroleum-based lubricants at sliding contacts in a way to help the environment and to reduce energy dissipation in industrial components for strategies toward energy efficiency and sustainability. In SLMMCs, solid lubricant materials including carbonous materials, molybdenum disulfide (MoS_2), and hexagonal boron nitride (h-BN) are embedded into the metal matrices as reinforcements to manufacture a novel material with an attractive self-lubricating properties. Due to their lubricious nature, these solid lubricant materials have attracted researchers to synthesize lightweight self-lubricating metal matrix composites with superior tribological properties. This chapter focuses on the recent development in tribological behavior of self-lubricating metal matrix (aluminum, copper, magnesium, and nickel) composites. It is important to note that the tribological parameters, such as normal load, sliding speed, and temperature vary on a wide range and also the counterface materials differ in different experimental tests, comparing the results of tribological behavior of different self-lubricating composites is extremely difficult. In this chapter, attempts have been made to summarize the tribological performance of various SLMMCs as a function of several tribological parameters. These parameters include material parameters (size, shape, volume fraction, and type of the reinforcements), mechanical parameters (normal load and sliding speed), and physical parameters (temperature and environment). The mechanisms involved for the improved mechanical and tribological performances are discussed.

E. Omrani · A. Dorri Moghadam · P.K. Rohatgi
Department of Materials Science and Engineering, College of Engineering & Applied Science, University of Wisconsin, Milwaukee, WI 53211, USA

P.L. Menezes (✉)
Department of Mechanical Engineering, University of Nevada, Reno, NV 89557, USA
e-mail: pmenezes@unr.edu

J.P. Davim (ed.), *Ecotribology*, Materials Forming, Machining and Tribology,
DOI 10.1007/978-3-319-24007-7_3

1 Introduction

Human being deplete and degrade natural resources during manufacturing of novel products which cause the earth confronting numerous serious environmental problems including global warming and environmental pollution which are advancing rapidly. Environmental issues, namely, global warming has reached a point at which action toward protection cannot be delayed. It unavoidably tends to critical demand for discover proper solutions in order to protect the environment. Hence, it is imperative to find innovative technologies to gain some degree of environmental sustainability.

One of the widespread multidisciplinary area for studying and conducting research of interacting surfaces in relative motion is tribology which mainly investigates the durability and efficiency of mechanical systems [1]. Therefore, integrating suitability ideas to this field of science will raise a good opportunity toward environmental problems. In this regard, environmental friendly tribology or ecotribology is focused on utilization of sustainable material and energy resources to save natural and nonrenewable resources and to reduce environmentally harmful effects [2].

Most of the commercially available lubricants destroy soil's micro flora. Likewise, waste lubricants are more harmful because they have slow process of biodegradation. Moreover, oil can penetrate into the drinking water supplies and very low concentrations in water make drinking water inadmissible to use due to the smell, taste, and health dangers. Besides, the ground and watercourses can consistently be contaminated by million tons of waste lubricants.

Saving natural resources can be attained by boosting the lifecycle of technical systems and recycling of used products. There are numerous fundamental tribological research emphasize the concern over the environmental issues. Taylor [3] investigated the significance of tribological design for components that are involved with wear condition in automotive internal combustion engines. In 1994, the number of registered vehicles was reported to be 500 million with an additional 40–50 million cars produced in the world every year. A rough estimate shows a reduction in the mechanical failure due to tribology is about 10 % that tends to decrease in the fuel consumption by 1.5 %. It means that the saving of about 340 L of petrol during life time of a car that leads to reduction in emission due to lower fuel consumption [3, 4]. In this manner, it becomes clear that if one integrate these small savings ($350) per car to the vast number of cars currently in use, the economic savings and environmental benefits are tremendous.

Several alternatives are conceivable to save energy resource and material, including optimum design, optimum operation, and optimum materials [5]. Replacing oil-based lubricants with solid lubricants is a promising approach to reduce environmental impacts and also to optimize materials design. Employing solid lubricants as coatings and reinforcements are well-studied techniques to eliminate the usage of oil-based lubricants [6, 7]. The impediments of using solid lubricants as coatings are limiting the applications of solid lubricants because of lifetime, difficulty in replacement, oxidation and aging-related degradation, and poor adhesion. Therefore, to avoid these disadvantages of the solid lubricant coatings, embedding solid lubricant into the metal matrix seems promising. Hence, self-lubricating composites, which is a new concept for materials design, were introduced as the prospective environmentally benign material for energy efficiency and sustainability [8].

2 Why Self-lubricating Materials?

In most tribological applications, liquid- or grease-based lubricants are used to facilitate the relative motion of solid bodies by minimizing friction and wear between interacting surfaces. A lubricant made of lower shear strength layer between two contacting surfaces and the shear strength of this layer is less than the surface shear strength between the sliding surfaces [9]. Therefore, this lower shear strength lubricant layer reduces friction between the surfaces during relative motion [10]. In fact, lubricants can separate the surfaces with no actual contact between two surfaces. This means that there is no formation of asperity junctions between the surfaces. However, in most cases, depending on thickness of the lubricant and testing conditions, asperities may have contacts and it is not possible to avoid asperity contacts completely; although lubricants are able to reduce it and may also reduce the shear strength of the junctions formed [11, 12].

The challenges for liquid lubricants arise in extreme environmental conditions, such as very high or low temperatures, vacuum, radiation, and extreme contact pressure. At these conditions, solid lubricants may be the alternative choice which can help to decrease friction and wear without incorporating liquid lubricants. Generally, when solid lubricants are introduced at the contact interface, solid lubricants function in the same way as that of liquid lubricants. They made of low shear strength layer that can shear easily between two surfaces and minimize direct contact between surfaces. Consequently, solid lubricants can provide low friction and diminish wear damage between the sliding surfaces. In addition, a mixture of solid and liquid lubrication is likewise achievable to have a beneficial synergistic effect to enhance the friction and wear performance. Solid lubricants can be dispersed in water, oils, and greases to attain improved friction and wear properties [13–21].

Several well-known inorganic materials have lubrication properties in nature and they are able to provide excellent tribological performance during sliding. These solid lubricants include molybdenum disulfide, carbonous allotropes, hexagonal boron nitride, and boric acid [13, 22–24]. The key feature of solid lubricants is that they have a lamellar or layered crystal structure that can provide adequate lubricity. Graphite, hexagonal boron nitride, boric acid demonstrates the layered crystal structures [25–29].

Challenges with solid lubricants are to maintain a continuous supply of solid lubricants on the contact surfaces to act as lubricous layer between two sliding surfaces. Such a continuous supply of solid lubricant is more easily maintained in the case of liquid lubricants when compared to solid lubricants. The most innovative development to ensure a continuous supply of solid lubricant to the contact surface during sliding is to introduce solid lubricant as reinforcement into the matrix of one of the sliding components. A self-lubricating material is one whose composition or structure facilitates low coefficients of friction and wear through the use of a self-dispensed and self-regulated lubricant delivery system, such as graphite, MoS_2, etc. These materials are becoming more and more attractive as the world become increasingly conscious of our environmental protection and energy usage. Self-lubricating composites have the potential to effectively increase our energy efficiency through more efficiently operating system components.

Metals, ceramics, and polymers are used as matrix materials to synthesize self-lubricating composites. Self-lubricating metal matrix composites (SLMMCs) can be processed by casting or powder metallurgy techniques [30–43]. Almost all metals and alloys are being researched to develop self-lubricating composites. Self-lubricating composites have been used for a long time and are utilized rather widely by the industry to combat friction and wear in a variety of sliding, rolling, and rotating bearing applications. Recent studies exhibit that some of wear particles produced at the interface are solid lubricant and they can form a thin film layer of solid lubricant on the contact surfaces of materials. This lubricious layer causes to decrease the friction coefficient and wear rate and enhance tribological properties. Therefore, composites reinforced by solid lubricant become self-lubricating due to the lubricant film developed at the interface, which prevents direct contact between the mating surfaces. Thus, self-lubricating composite eliminates usage of any types of external lubricants by reducing friction and wear due to self-lubricating nature of the materials. This lubricant film initially does not present and it forms as a result of worn surface and subsurface deformation. They are continuously replenished by embedded solid lubricant particles in the matrix [44, 45]. For example, aluminum/graphite composites show an improvement in lubricity, durability, and resistance to seizure under both dry and lubricated conditions [45].

Similar to metal matrix composites, polymer and ceramic matrix composites are favorable materials for several industries due to the superior properties [46–53]. It is possible to embed solid lubricants into polymer and ceramic matrix and fabricate self-lubricating polymer [54] and ceramic matrix composites [55]. In this chapter, self-lubricating behavior of various solid lubricants-reinforced metal matrix composites is reviewed. More specifically, the tribological properties of aluminum, copper, magnesium, and nickel matrix composites have been discussed. The influence of the matrix alloy composition, the solid lubricant content and its shape and size [56–59], and the testing parameters, such as contact load [60–63], sliding velocity [60–64], temperature [55, 65–67] on the tribological performance of metal matrix composites are discussed. The mechanisms of solid lubricant film formation on the tribo-surfaces of these composites were also discussed as the superior tribological properties of metal matrix composites are strongly dependent on the formation of lubricant film on the tribo-surfaces.

3 Self-lubricating Metal Composites

Man-made self-lubricating materials primarily involve the creation of some type of composite materials. A composite is a coupling of two different materials designed to inherent the qualities of both materials. Self-lubricating composites take an advantage of a hard structural matrix carefully combined with a lubricating phase. There are several ways to incorporate the lubricating phase. Dispersing solid lubricant particles or fibers throughout the matrix can be a simple and an effective way to ensure that the material is constantly lubricated. The properties of the individual matrix and lubricant, concentration of the lubricating phase, distribution or order of the lubricating phase, and interactions between the lubricant and the matrix are variables that determine the quality of these types of composites. A schematic of self-lubricating composite is shown in Fig. 1. As shown in the figure, the material wears against the contact surface, new solid lubricant particles will be exposed to the surface thereby keeping the surface lubricated. A classic example of this type of composite is the gray cast iron; it utilizes a hard iron matrix with dispersed lubricating graphite flakes. Constructing a composite with alternating layers of the structural phase and the lubricating phase is also an effective way to engineer self-lubricating composites.

As mentioned earlier, the distinctive feature of self-lubricating composites is that the wear particles formed on the contact surface acts as solid lubricants and it can reduce the friction coefficient and wear rate. For instance, under sliding conditions, the metal/graphite composite can form self-lubricating composite because of the transfer layer formation of the graphite on the tribo-surfaces during sliding and this transfer layer acts as a solid lubricant film which prevents direct contact between the mating surfaces [26]. To have an effective lubricant layer, it is also important that the solid lubricant has a strong adhesion on the bearing surface; otherwise, this lubricant layer can be easily rubbed away and tends to very short service life.

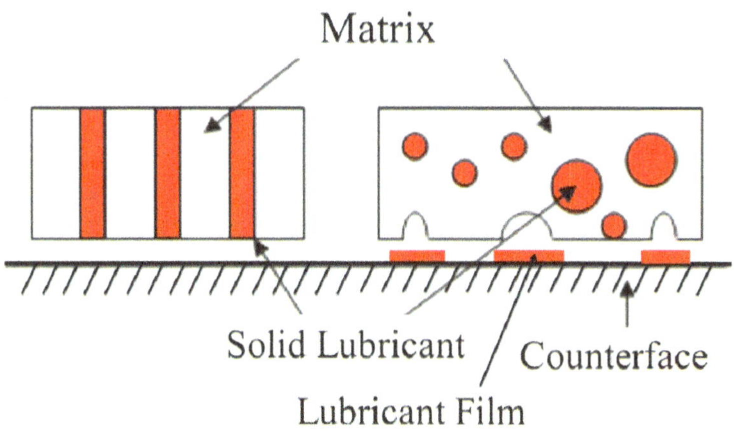

Fig. 1 Schematic of self-lubricating composite and its mechanism

3.1 Aluminum Matrix Composites

With increasing demands for high-performance materials for use in energy efficient, low maintenance engineering systems, lightweight and the high-strength aluminum alloys are of great interest to designers [68]. Excellent properties of aluminum alloys and aluminum matrix composites (AMC), such as high specific strength [69], high corrosion resistance [70], good thermal conductivity [71], low electrical resistivity [72], and high damping capacity [73] tend to increase their applications in several industries, including aerospace, marine and automotive for components, such as engines, cylinder blocks, pistons, and piston insert rings [13, 74, 75]. Furthermore, these materials make a good substitution for cast iron because of poor properties including low corrosion resistivity and low strength/weight ratio [76–78]. The best substitution for cast iron and bronze alloys is the AMCs when tribology is the main concern because of their superior wear and seizure resistance [79, 80].

Generally, aluminum matrix reinforced with graphite particles exhibit improved tribological properties in comparison with aluminum composite reinforced with other ceramic particles, such as Al_2O_3 and SiC. The friction and wear rate decrease significantly in the presence of solid lubricant particles as reinforcements when compared to unreinforced matrix alloys as a result of incorporation of graphite particles [81]. Ames et al. [82] investigated the effect of graphite on wear regime of composites. The results showed that the composites reinforced by graphite continued to remain in the mild wear regime even at high normal load conditions and did not exhibit any severe wear, while its alloy observed the severe wear regime at high normal loads (Fig. 2). The reason for this behavior is the formation of tribofilm during sliding that can provide adequate lubrication between contact surfaces.

Fig. 2 Effect of graphite particles on transition point of metal matrix composites [82]

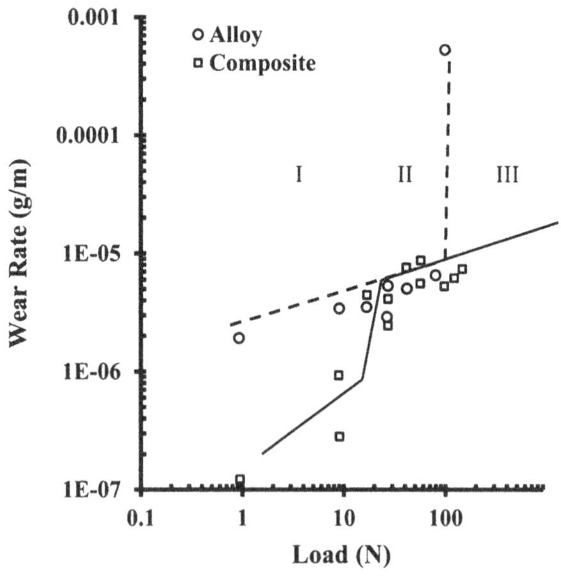

Important factors that can affect tribological properties are type [22, 56, 83], size [22, 56, 57, 84], volume fraction [22, 56–58, 85], and distribution [86] of the reinforcing phase as well as the manufacturing technique of the composites [32, 87, 88]. Another important factor that can affect the mechanical properties and tribological behavior of aluminum/graphite composites is the interfacial bonding between the matrix and graphite [89, 90]. Furthermore, testing parameters, such as load and sliding velocity can affect the tribological properties of self-lubricating aluminum/graphite composites [22, 61–63].

The effect of volume fraction of graphite content on the COF and wear rate is shown in Fig. 3a, b, respectively. As shown in Fig. 3, there is a significant influence of graphite content on the tribological properties where the friction coefficient and wear rate of aluminum/graphite self-lubricating composites decreases with increasing the amount of graphite content [38, 88, 91–94] due to increasing the thickness of the lubricating film at the interfaces with increasing graphite content,

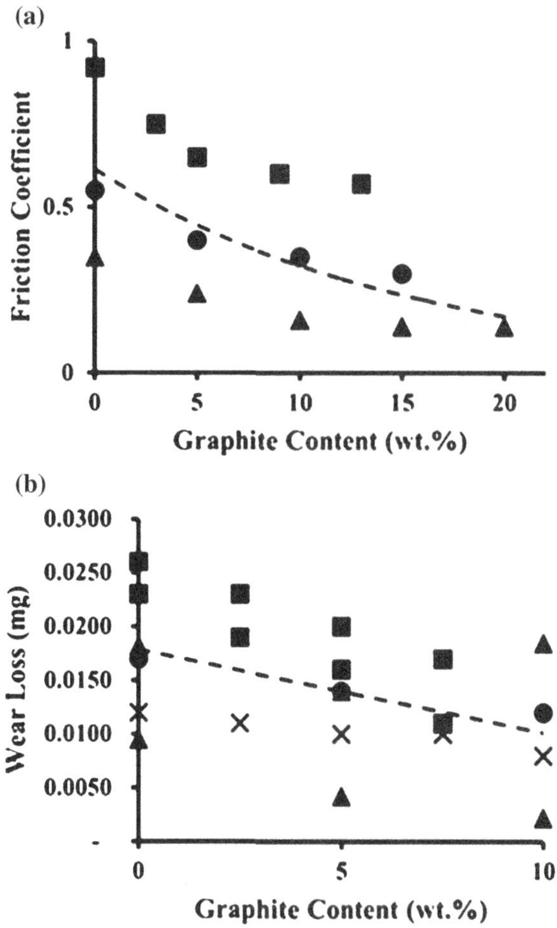

Fig. 3 Correlation between graphite content and **a** friction coefficient [41, 42, 88] **b** wear loss [38, 91, 92]

which ultimately decreases the asperities contacts [94, 95]. This phenomenon occurs as a consequence of the shearing of graphite particles employed between the sliding surfaces of the composite and counterpart material. It reduces the magnitude of shear stress and plastic deformation in the subsurface region as well as minimizing direct metal-to-metal contact. This lubricant layer also acts as a solid lubricant between the two sliding surfaces [96].

Figure 4 depicts the worn surface of unreinforced aluminum and aluminum reinforced with graphite. When comparing worn surfaces of unreinforced aluminum alloy (Fig. 4a) with composite reinforced by 5, 10, and 20 wt% graphite (Fig. 4b–d), it can be expected that the grooves are less on composites than the unreinforced alloy. Deep abrasive grooves were formed on sliding surface as a result of severe plastic deformation. Additionally, surface of composites shows graphite layer that the thickness of tribofilm increases with increasing graphite volume fraction in the

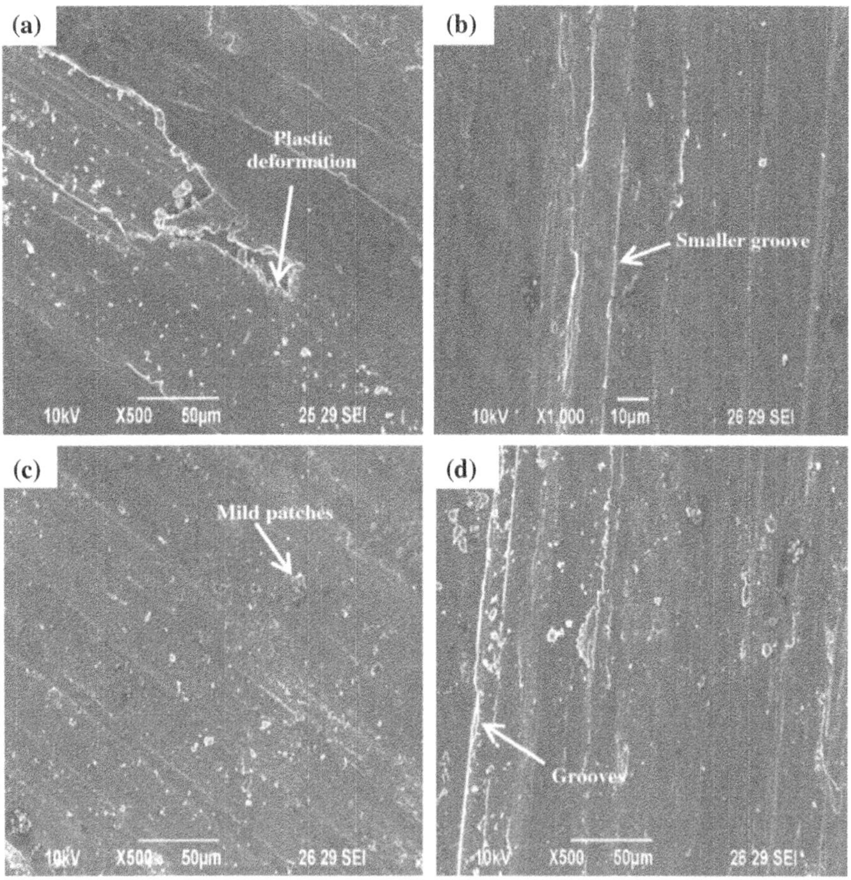

Fig. 4 SEM micrographs of worn surface of **a** unreinforced aluminum A7075 **b** A7075/5 wt% graphite **c** A7075/10 wt% graphite **d** A7075/20 wt% graphite [96]

composite. This graphite tribolayer act as a protective layer to preclude direct contact between the composite surface and counter surface. Thus, graphite layer on the surface can effectively reduce the friction coefficient of the metal matrix composite reinforced by graphite particles [96, 97].

Generally, the friction coefficient and wear rate of aluminum/graphite composites increase with increasing applied normal loads as shown in Fig. 5a, b [38, 64, 91, 93–95, 98–102]. At higher normal loads, due to severe plastic deformation, the wear rate increases and hence delamination wear occurs [91]. Dominant wear mechanism at low loads is abrasion that can be inferred due to the presence of grooves formed by the reinforced particles on the worn surfaces of the composites along the sliding direction. With increasing normal loads, the grooves on the contact surface become deeper, consequently, delamination becomes the dominant wear mechanism at higher normal loads due to the fracture strength of the particles are less than applied stresses [98]. Accordingly, at very high loads, the plastic flow of the material becomes dominant. Extensive plastic deformation causes severe

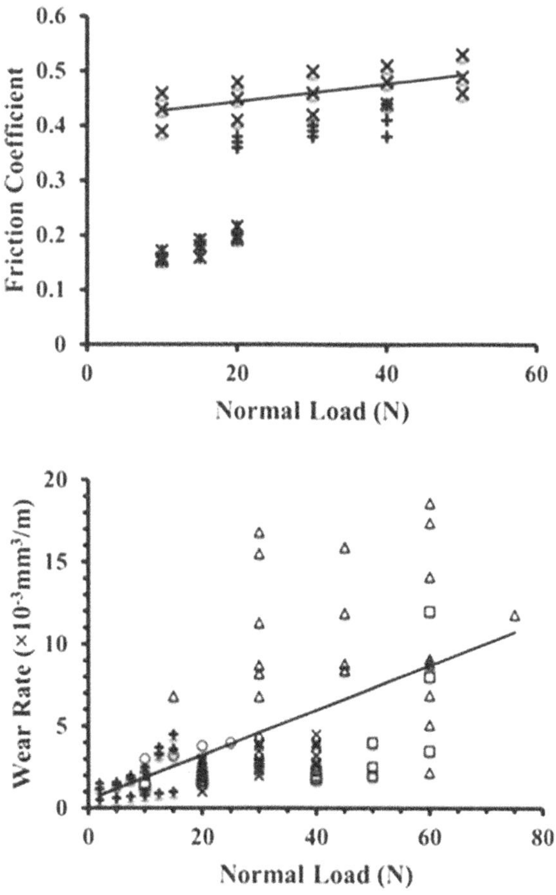

Fig. 5 Correlation between applied load and **a** friction coefficient [38, 94, 100] and **b** wear rate [64, 93, 94, 98, 100]

wear, and that is the key reason for severe wear conditions occurring at higher loads [100]. Generally, parallel ploughing grooves and scratches can be seen all over the surfaces along the sliding direction [94].

For a constant volume fraction, it is generally observed that when reinforcement particle size decreases, it influences strength, ductility, machinability, and fracture toughness [103–106]. Even though, a few studies have investigated the effect of graphite particle size on tribological properties, and its effect on mechanical properties has been explored extensively [91, 96, 107–110]. Jinfeng et al. [111] investigated the effect of different particles size (average diameter: 1, 6, and 20 μm) at a constant volume fraction on wear rate of Al/SiC/Gr hybrid composites. As shown in Fig. 6, it has been demonstrated that the wear loss of Al/SiC/Gr composites gradually reduced with increasing graphite particle size, from 2.7 mg wear loss for graphite particle size of 1 μm–1.4 mg wear loss for graphite particle size of 20 μm. Thus, Al/SiC/Gr composites with finer graphite particles exhibited lower wear resistance [111].

The greatest disadvantage of aluminum/graphite self-lubricating composites is the low mechanical properties. By increasing the amount of graphite in the matrix, the mechanical properties of AMC reinforced by graphite decreases [41, 92, 96]. Many approaches have been employed to reduce the damaging effect of graphite particles on the deterioration of mechanical properties. Two well-known methods—(a) manufacturing hybrid aluminum metal matrix composites containing ceramic particles and graphite particles and (b) embedding nanosized carbounous materials,

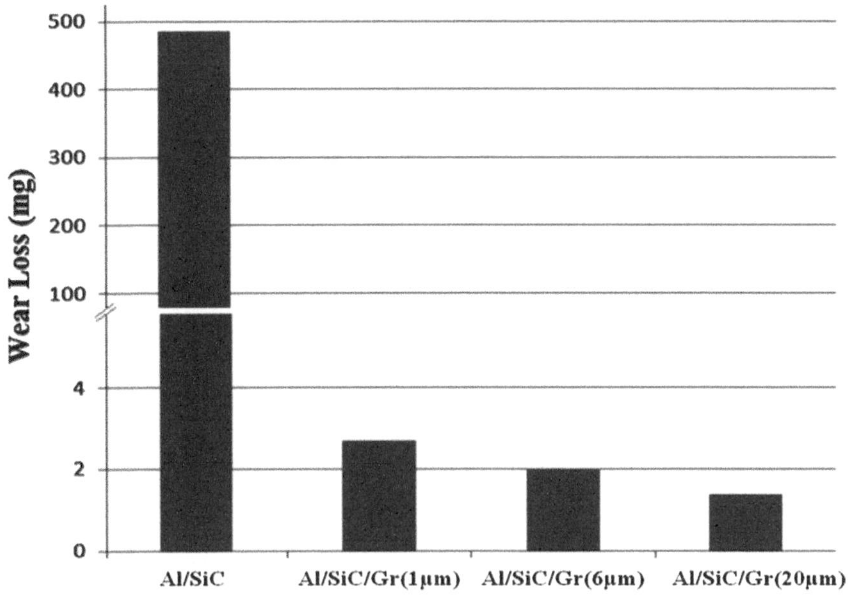

Fig. 6 Wear loss of the Al/SiC and Al/SiC/Gr composites [111]

such as carbon nanotubes (CNTs) and nanographite or graphene [22] in aluminum metal matrix—in order to achieve improved mechanical, electrical, and tribological properties.

Previous studies revealed excellent tribological properties of metal/CNTs composites due to the lubricating nature of CNTs. Akin to graphite, CNTs can form a lubricant film between contact surfaces during sliding and reduce direct asperity contacts. There is a week van der Waals bonds between CNT-metal interface which is directed to effortlessly slide or roll between the sliding surfaces and diminish a direct contact between the surfaces. Consequently, composites reinforced by CNTs show a decrease in friction coefficient of the composite. The enhancement in wear resistance is due to the character of CNTs as spacers between contact surfaces that avoid direct contact between asperities of rough surfaces [112]. Generally, several material parameters, such as amount of reinforcements, size of reinforcement, and spatial distribution have a direct effect on tribological properties of self-lubricating metal/CNTs composites [22, 113].

Zhou et al. [114] have investigated the mechanical and tribological properties of Al–Mg/Multiwall Carbon Nanotubes (MWCNT). As shown in Fig. 7a, addition of MWCNT into Al–Mg alloy improved the hardness of the composites compared to unreinforced aluminum alloy. With increasing MWCNTs volume percentage into the matrix, the hardness of composite initially increased and then decreased at higher amount of MWCNTs. The effect of volume fraction of MWCNTs on the friction coefficient and wear rate of the composite is shown in Fig. 7b. The coefficient of friction and wear rate decreased even at high MWCNTs volume fraction. The X-ray diffraction (*XRD*) analysis of contact surface revealed that the wear particles are mainly aluminum oxide. During wear process, laminated oxide films were formed at contact surfaces that they subsequently broke up and flaked off due to low adhesion between the oxide films and aluminum matrix. While the oxide particles which are present at contact surface are harder than the aluminum matrix, and hence they are able to increase abrasive wear. As the aluminum matrix gradually wear out during sliding process, the CNTs which were initially embedded in the matrix are now pulled out and exposed to the contact surface and form a lubricant film on the worn surface. This solid lubricating film significantly reduces the abrasive wear cause by oxide particles compared to unreinforced aluminum.

Choi et al. [115] have investigated the effect of test parameters, including normal load and sliding speed on coefficient of friction and wear rate. Figure 8 demonstrates the effect of applied load and sliding velocity on coefficient of friction and wear loss. Investigations have revealed that the COF and wear loss increases with increasing normal load for aluminum/4.5 vol% MWNT composite at constant sliding speed of 0.12 m/s. Nevertheless, the coefficient of friction is still lower than 0.1. At higher applied load, the friction coefficient and wear loss are increased as severe wear plays the dominant wear mechanism and as a result, severe surface damage was observed at higher applied loads. On the other hand, the coefficient of friction and wear loss has slightly decreased with increasing sliding speed at a constant applied load of 30 N.

Fig. 7 Variations of **a** Brinell hardness (HB) **b** wear rate and coefficient of friction with MWNTs content for melt-infiltrated MWNT/Al–Mg composites under an applied load of 30 N and a sliding velocity of 1.57 m/s [114]

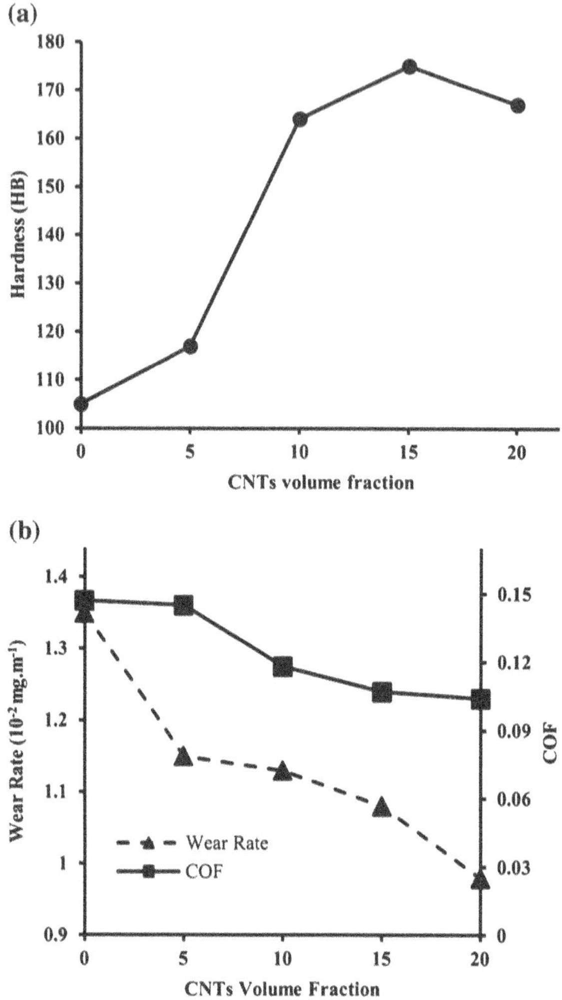

Ghazaly et al. [116] synthesized AA2124/graphene nanocomposites and investigated the effect of weight percentage (0.5, 3, and 5 wt%) of graphene on mechanical and tribological properties. As shown in Fig. 9, the self-lubricating composite reinforced by 3 wt% graphene exhibited the best tribological properties under dry wear test in comparison with unreinforced and other nanocomposites. In all samples, longitudinal grooves were observed on worn surfaces of unreinforced aluminum alloy and aluminum/graphene nanocomposites. Furthermore, size of scratches, craters, and delamination of AA2124/3 wt% graphene composite is smaller than the unreinforced aluminum alloy and aluminum/graphene nanocomposites. The wear mechanism of AA2124 was severe wear while the wear mechanism was changed to mild wear for AA2124/3 wt% graphene nanocomposite.

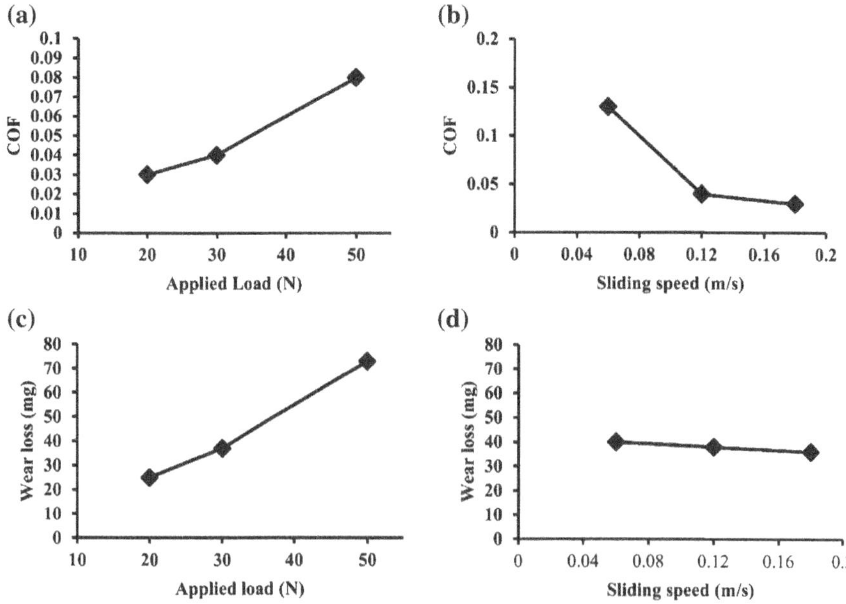

Fig. 8 Variations of COF of 4.5 vol% MWNT/Al composite with **a** applied load at sliding speed 0.12 m/s and **b** sliding speed at applied load 30 N. Variations of wear loss of 4.5 vol% MWNT/Al composite with **c** applied load at sliding speed 0.12 m/s and **d** sliding speed at applied load 30 N [115]

Fig. 9 Wear rate and weight loss variation as a function of graphene content in AA2124 matrices [116]

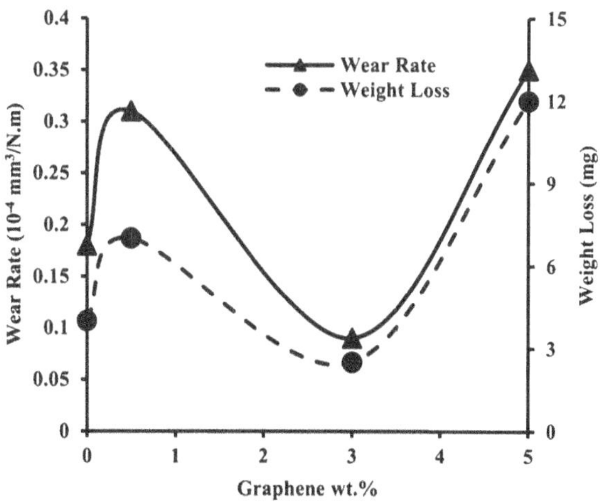

Shallow parallel grooves and ridges were formed on the worn surface of AA2124/0.5 and 5 wt% graphene nanocomposite due to micro-ploughing. Thus, the dominant wear mechanism was severe plastic deformation of the matrix that results in a high wear rate. At higher magnification, the AA2124 alloy had debris on the worn surface while there is no wear debris on the worn surfaces of nanocomposite as illustrated in Fig. 10. Alumina fragmented films or strain-hardened particles were the two main sources for debris formation. Detached consolidated powders were formed as wear debris on the contact surfaces. By comparing the worn surfaces at high magnifications, the nanocomposite containing 3 wt% graphene exhibited smoother surface against other counter materials. Additionally, AA2124/3 wt% graphene composite was covered by lubricant films on the surface

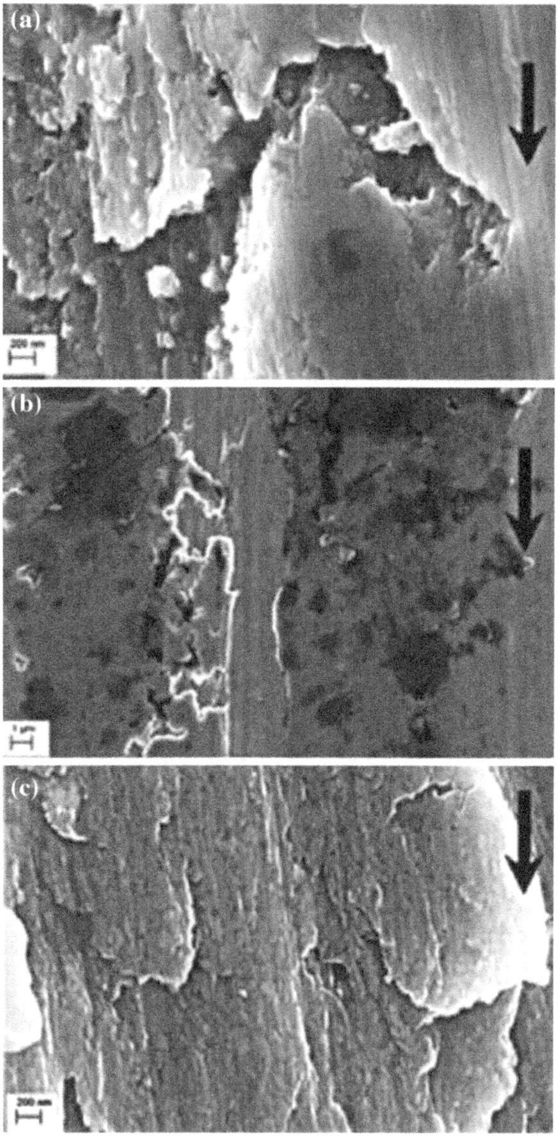

Fig. 10 SEM micrographs for the worn surfaces of AA2124—**a** 0, **b** 3 and **c** 5 wt % graphene nanocomposite [116]

and as a result it tends to decrease friction and wear rate due to the soft nature of the lubricant film. Therefore, shallow grooves and mild damage occur on the worn surfaces of AA2124/3 wt% graphene composites along the direction of sliding, while surface of AA2124/5 wt% graphene exhibited deep grooves and severe damage. Hence, a significant increase in wear rates and weight loss for AA2124/5 wt% graphene was observed.

Zamzam [117] had synthesized aluminum/3graphite, aluminum/5MoS_2, and aluminum/3graphite/2MoS_2 composites. The samples were tested in two different conditions: as-extruded and annealed (813 K for 6 h). Figure 11 compares the weight loss of different composites reinforced by different solid lubricants as well as the effect of annealing on weight loss of aluminum/3graphite/2MoS_2 composites. The as-extruded samples show lower weight loss than those tested after annealing. The reason for the negative effect of annealing is that the oxidation of MoS_2 occurs at 573 K and generate MoO_3, while the oxidation of graphite occurs at 723 K and produce CO and CO_2. Therefore, the annealing treatment eliminated the lubricity nature of MoS_2 and graphite and resulted in higher wear loss. Furthermore, the results revealed that aluminum reinforced by graphite possesses better wear resistance than reinforced by MoS_2. In addition, aluminum/MoS_2 and aluminum/graphite/MoS_2 show a transition point from mild wear to severe wear at 8 and 38 km, respectively, while aluminum/graphite did not exhibit any failure and severe wear up to 60 km. This phenomenon could be attributed to variations in temperatures at the contact surface. Thus, temperature of contact surface is more than oxidation temperature of MoS_2 and less than oxidation temperature of graphite and it tends to oxidize of MoS_2. When oxidation occurred, severe wear phenomena initiated and as a result failure started to observe on the surfaces. For composites reinforced by graphite, oxidation never occurred as the surface temperature is less than oxidation temperature of graphite. Consequently, it means more stability for

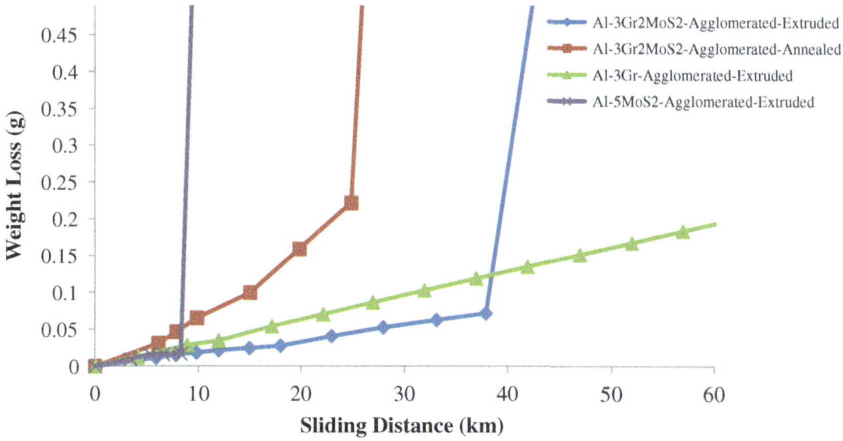

Fig. 11 Effect of types of solid lubricant on weight loss of aluminum composites [117]

graphite on the contact surface than MoS_2 and thus delayed the onset of failure of composites reinforced by graphite and graphite did not show any transition.

Dharmalingam et al. [118] studied the tribological properties of Al-Si10Mg matrix reinforced by alumina (10–20 μm) as hard reinforcement and molybdenum disulfide (1.5 μm) as soft reinforcement. The volume fraction of alumina was 5 wt% and molybdenum disulfide was varying between 2 and 4 wt%. The effect of volume fraction of MoS_2, applied load, and sliding speed on the tribological performance were investigated. As seen in Fig. 12 that the lowest specific wear rate was observed in the composite reinforced by 4 wt% of molybdenum disulfide at 30 N applied load and 4 m/s sliding speed. The lowest coefficient of friction was obtained for composite reinforced by 4 wt% of molybdenum disulfide at 10 N applied load and 4 m/s sliding speed as shown in Fig. 13. The results was explained by the SEM micrographs of the worn surface of aluminum composite specimens. The composite reinforced by 4 wt% molybdenum disulfide clearly exhibited fine grooves with least amount of plastic deformations at 10 N applied load and 4 m/s sliding speed as shown in Fig. 14a. In contrast, at higher applied load (50 N) for the same sliding speed (4 m/s), grooves on the worn surface of the composites reinforced by 4 wt% molybdenum disulfide became deeper and wider, besides, the amount of plastic deformation at the edges of the grooves were high as shown in Fig. 14b. Many researchers stated that the oxidative wear was the dominant wear mechanism that occurs during dry sliding wear of aluminum composites. The amount of oxide layer as lubricant film improved with the increase in the amount of molybdenum disulfide at the interface suggesting that the worn out surfaces were oxidized. In

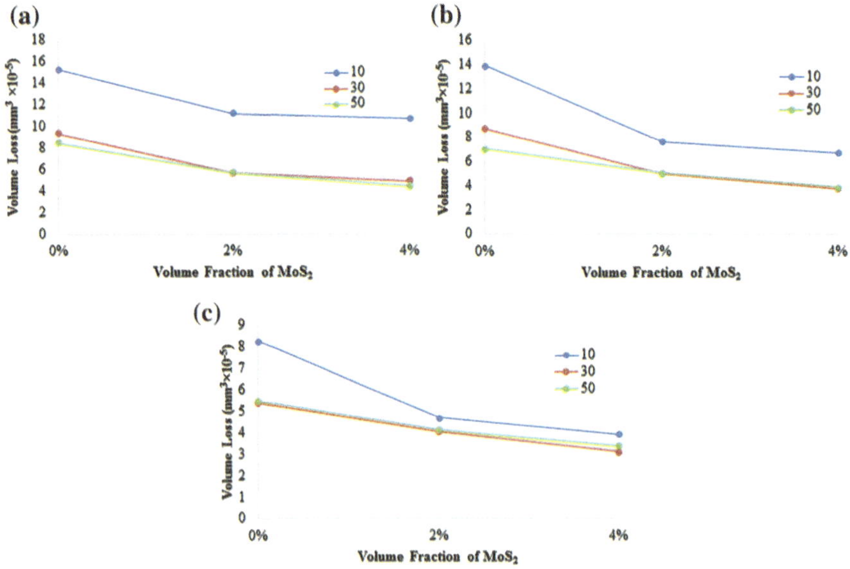

Fig. 12 Wear rate of Al-Si10Mg/Al_2O_3/MoS_2 composite at sliding speed of **a** 2 m/s **b** 3 m/s **c** 4 m/s [118]

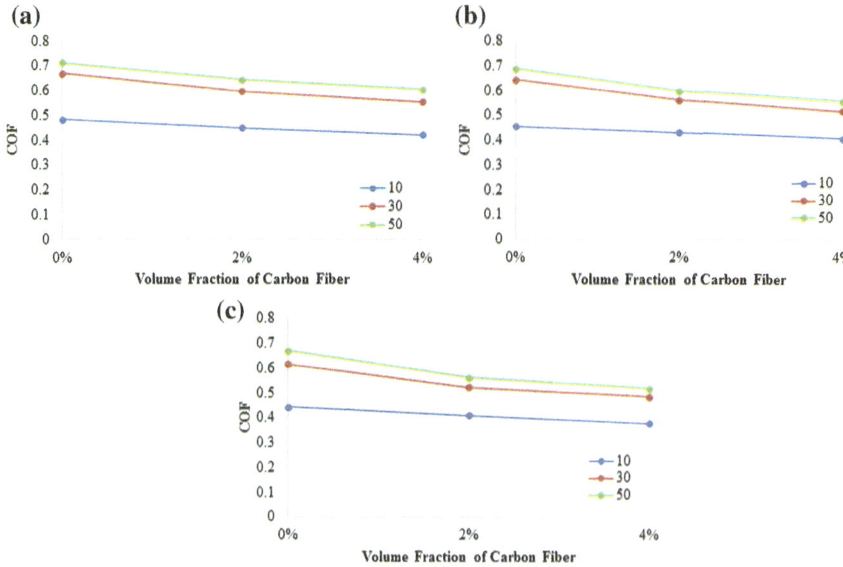

Fig. 13 Wear rate of Al-Si10Mg/Al_2O_3/MoS_2 at sliding speed of **a** 2 m/s **b** 3 m/s **c** 4 m/s

aluminum-alumina-molybdenum disulfide composites, higher amount of oxide was formed at higher sliding speed for a constant applied load, which leads metal-to-molybdenum disulfide contact is increased, therefore there is less metal-to-metal contact and consequently, lowering the specific wear rate.

3.2 *Copper Matrix Composites*

Advantages of copper composites reinforced by solid lubricants are excellent thermal and electrical conductivities and low thermal expansion coefficient as they can retain the properties of copper matrix materials. Because of these excellent properties and self-lubricating nature of copper/solid lubricant composites, they have been extensively used as contact brushes and bearing materials in many industrial applications. In the case of low voltage and high current densities, typically for sliding parts of welding machines, it is required to employ materials with a very high specific electrical conductivity, good thermal conductivity, and low friction coefficient. Such conditions are fulfilled only by using copper/graphite composite materials. Copper matrix containing graphite composites are widely used as brushes, and bearing materials in many applications.

Moustafa et al. [119] investigated the effect of graphite content (8, 15, and 20 wt %) at different normal loads (50–500 N) on the tribological performance of the composites synthesized by powder metallurgy. They used either Cu-coated graphite powders or mixture of copper and graphite powders. The pin-on-ring tribometer

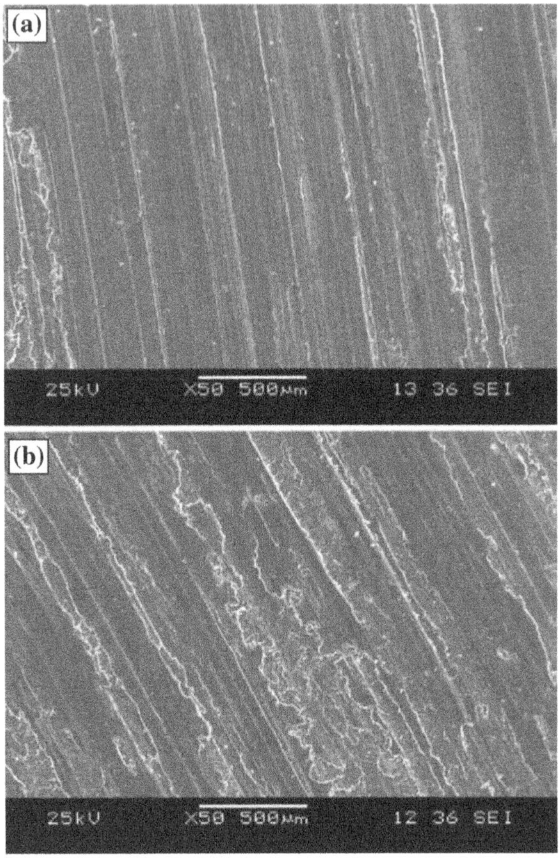

Fig. 14 SEM micrograph of worn surface of aluminum-5 wt% of alumina-4 wt% of molybdenum disulfide at 4 m/s. **a** 10 N and **b** 50 N [118]

was employed for the wear testing. Figure 15 shows the comparison of volumetric wear rate for copper/graphite composites at different normal loads. The copper/graphite composite showed lower wear rate than the sintered copper compacts up to 200 N normal loads due to the presence of a smeared graphite layer on the contact surface of the wear sample which is produced by extrusion of graphite between the surface of the tested pin during sliding. This lubricant layer acts as a solid lubricant and it tends to reduce the wear rate. Composites containing 8 and 15 % graphite can endure the highest loads up to 450 N, while the composite containing 20 % graphite can withstand up to 500 N applied normal load. The coated graphite composites possess lower wear rate than that of uncoated composites. Figure 16 presents the variation of coefficient of friction at different applied loads for pure copper, coated and uncoated self-lubricating copper/graphite composites. The influence of graphite is significant on the coefficient of friction due to the presence of the graphite layer at the sliding surface of the wear sample, which acts as a solid lubricant. In addition, the coated graphite composites possess lower coefficient of friction than that of uncoated composites. Results revealed reduction

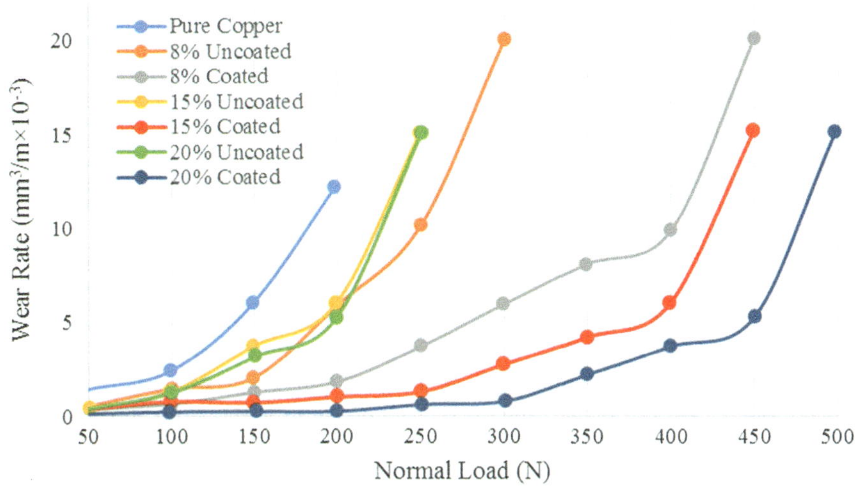

Fig. 15 Variation of wear rate with applied normal load for pure copper and Cu-coated and uncoated graphite composites with graphite contents of 8, 15, and 20 % [119]

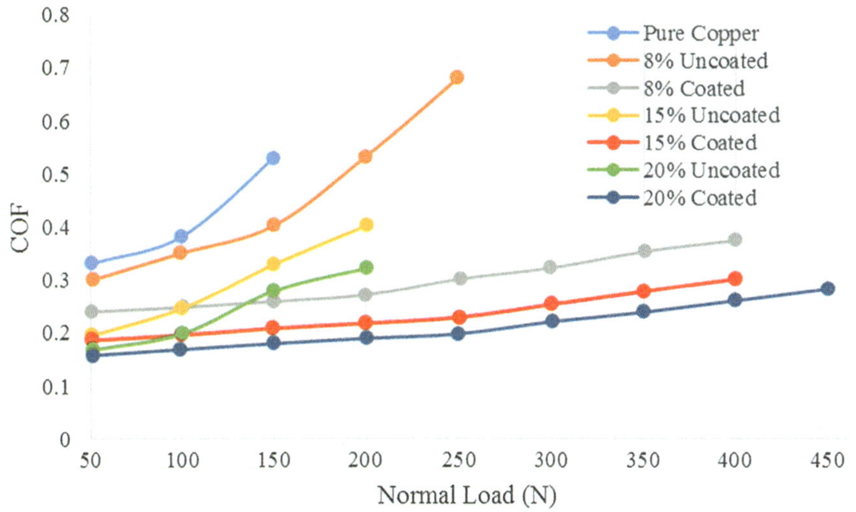

Fig. 16 Variation of coefficient of friction with applied load for: pure copper and Cu-coated and uncoated graphite composites of graphite contents of 8, 15, and 20 % [119]

in the coefficient of friction by increasing graphite content in either cases of coated and uncoated composites.

Thicker and denser lubricant film was observed in higher amounts of graphite-reinforced composites. Increase of volume fraction of the graphite content in the coated composites tends to significant reduction in wear rate for all graphite contents. Results showed that coated composites become high adherent and compacted than uncoated composites because of the high mismatch and good contact between graphite and copper matrix. Accordingly, weak bond between the graphite and copper matrix leads to the fast removal and regeneration of a layer during sliding wear test. Hence, copper/uncoated graphite composites have higher wear rates and coefficient of friction than copper/coated graphite at each corresponding graphite volume fraction.

Zhao et al. [120] investigated the effect of graphite reinforcement on copper matrix and tried to find the wear mechanism by SEM and EDX methods. The results of COF and wear rate are shown in Fig. 17. The friction coefficient and wear rate of self-lubricating copper composites decreased with increasing volume fraction of graphite. The slippage between the layers of graphite proceeds easily at a test load because of the hexagonal structure. Graphite sticks to the worn surface and it tends to form a solid self-lubricating layer on the worn surface. Thus, it leads to greatly improvement in wear properties of Cu–graphite composites in comparison with that of pure copper due to reducing direct contact between two surfaces and transforming contacts between metal and metal into the contacts between graphite film and metal or graphite film and graphite film.

Figure 18 shows the SEM and EDX analysis of worn surface of pure copper and Cu–graphite composites. When pure copper sample contacts with the counter material, adhesion occurs between the asperities, while less adhesion occurs when copper/graphite contacts with the counter material and results in less split fragments. Adhesion hinders the easy sliding between the pure copper against counter material, which results in a large friction coefficient and wear mass loss. Near the adhesion areas, plastic deformation was observed and microcracks and voids are formed at the surface (Fig. 18a) [121] while less plastic deformation and ploughing

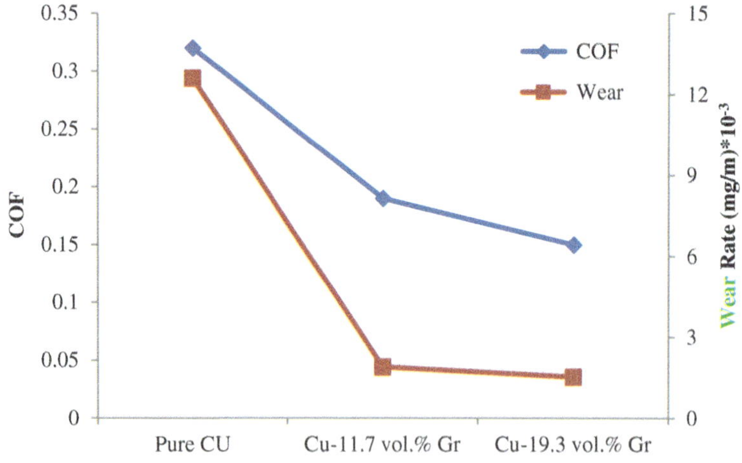

Fig. 17 COF and wear rate of Cu and Cu–graphite composites [120]

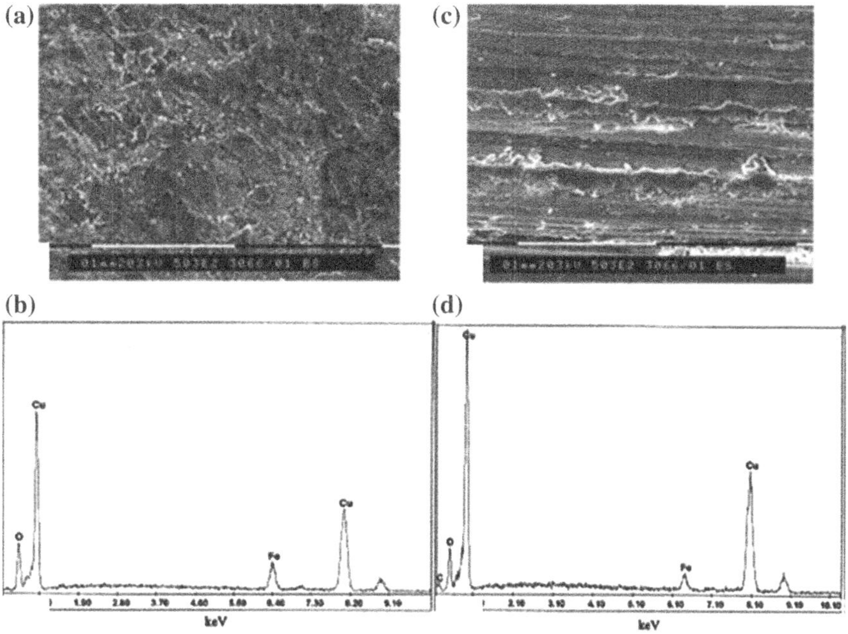

Fig. 18 **a** SEM micrograph and **b** EDX analysis on worn surface of pure copper **c** SEM micrograph and **d** EDX analysis on worn surface of Cu/graphite [120]

grooves along the sliding direction are detected on the worn surface of Cu/graphite composites as shown in Fig. 18c. The microcracks grow and extend to the free surface of pure copper results in debris formation, and likewise many copper debris produce on the worn surface as shown in Fig. 18a. On the other hand, the graphite film formed on the surface causes to change contact areas between the counter material and copper matrix to the graphite film and counter material. Besides, the graphite film has weak bonding between its layers and it leads to distribute the lubricant film on the worn surface and the counter material. Accordingly, for copper/graphite composite, the wear mechanism is changed into delaminating wear. In addition, some copper debris present on the surface of the counter material in the case of pure copper and there is no obvious transfer film of the pure copper on counter material. Furthermore, mild oxidative wear occurs in terms of "O" content as shown in Fig. 18b for pure copper; on the other hand, the amount of Fe and O on the worn surface of the composites decreased in comparison with pure copper as revealed in EDX analysis as illustrated in Fig. 18d. Consequently, graphite particles can enhance the tribological properties and reduce oxidative wear.

Similarly, several researchers showed improvement in tribological properties of copper matrix embedded by graphite particles. Kovacik et al. [113] investigated the effect of graphite at high volume fraction of up to 50 vol%. As graphite has good lubricating properties, embedding small volume fraction of graphite into copper matrix composite tends to significant improvement in coefficient of friction of copper composites up to a critical point. It was confirmed that there is a critical point for the graphite concentration. Below this critical point, the coefficient of friction and wear rate of the composites decreases with increasing concentration of graphite, while above this critical point, the coefficient of friction of the composites becomes independent on the volume fraction of graphite (Fig. 19). For metal matrix composites, the critical point is not about 20 vol% of graphite [45]. However, it significantly depends on the matrix and reinforcement. For example, it is 12 vol% for uncoated graphite with fine graphite powder (16 μm) while 23 vol% for uncoated graphite with coarse powder (25–40 μm) [119]. For coated graphite, the critical point was reported to be above 25 vol% of graphite. The reason is that the discontinuous graphite film is formed on the surface at low graphite volume fraction and then with increasing volume fraction of graphite, the graphite film becomes less separated and further increasing the amount of graphite, there is a homogenous graphite film that covers the complete contact area of composite material. This tends to form a graphite-rich mechanically mixed layer (MML) between the contacting surfaces. A large amount of graphite in the MML can reduce the shear strength in the near-surface region of the contact, thus decreasing the coefficient of friction and wear rate. By comparing uncoated and coated graphite, it is apparent that the coated graphite avoids bonding between graphite particles and reduces their clustering. Consequently, graphite phase has smaller size and smaller average distance between particles. For this reason, the coefficient of friction of coated composites decreases significantly at lower volume fraction of graphite in comparison with uncoated composites.

Figure 20 shows the effect of load on the coefficient of friction of copper and its composites. It is clearly confirmed that the coefficient of friction gradually decreases with increasing loads up to a load of 30 N. Beyond this applied load, the coefficient of friction of all composites gradually remains almost constant. The reason is due to

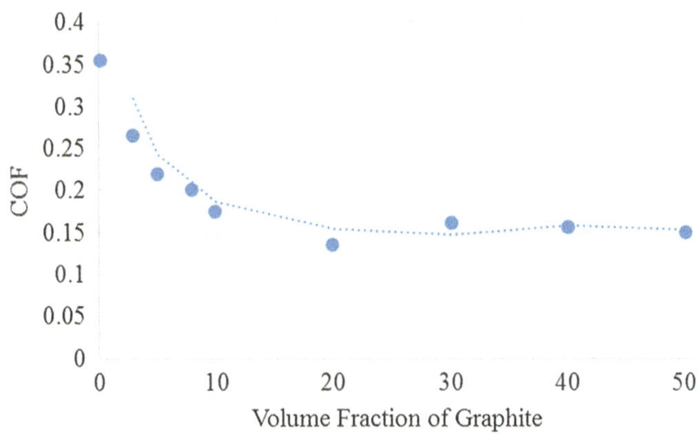

Fig. 19 Volume fraction dependence of friction coefficient of uncoated Cu–graphite composites at 100 N [113]

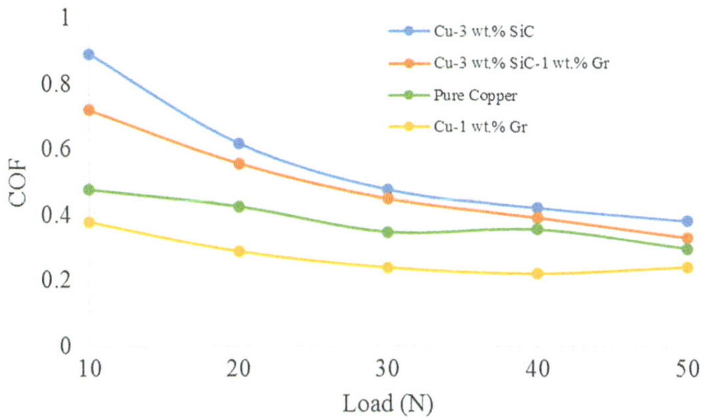

Fig. 20 Variation of coefficient of friction of copper and copper–SiC–graphite with load [122]

the fact that the copper matrix becomes soft and highly ductile under severe deformation. Consequently, it results in easy transfer of the copper metallic film on the counter material and this film prevents direct contact between the specimen and counter material leading to lowering of coefficient of friction. Additionally, copper/graphite composites exhibited the lowest coefficient of friction and copper/silicon carbide composite exhibited the highest coefficient of friction, while Cu–SiC–graphite composite showed an intermediate value.

The presence of graphite reinforcement has a negative effect on the mechanical properties of copper composites. The best choice is by employing nanosized particles to solve the negative impact of microsized graphite particles. Rajkumar et al. [30] have investigated the tribological properties of copper nanocomposite reinforced by 5–20 vol% nanographite (NG) particles with an average particle size of 35 nm. The variation of normal load with coefficient of friction at different volume percentage of micro- and nanographite particles is shown in Fig. 21. Figure 22 shows the variation of wear rate with normal load for copper-based composites reinforced by micro- and nanographite particles. It can be inferred from Figs. 21 and 22 that the coefficient of friction and wear rate both increase with increasing applied load. Results revealed that, at constant 15 % volume fraction, nanocomposites have better COF compared to the composite reinforced by micron-sized graphite particles due to higher hardness, lower porosity, and finer microstructure. Therefore, the nanographite particles-reinforced composites are more effective to the degree of self-lubrication compared to micron-size graphite particles-reinforced composites. In addition, volume fraction of nanographite can affect the tribological properties of self-lubricating copper composites. Increasing the amount of nanographite tends to decrease the COF up to 20 %, and thereafter the COF increases. The reason for decreasing of COF and wear rate at high volume fraction of nanographite is formation of lubricant film with more availability and uniformity. In general, the lubricant film tends to decrease the metal-to-metal contacts between the copper matrix composite and steel counter surface. Conversely, at higher amount of

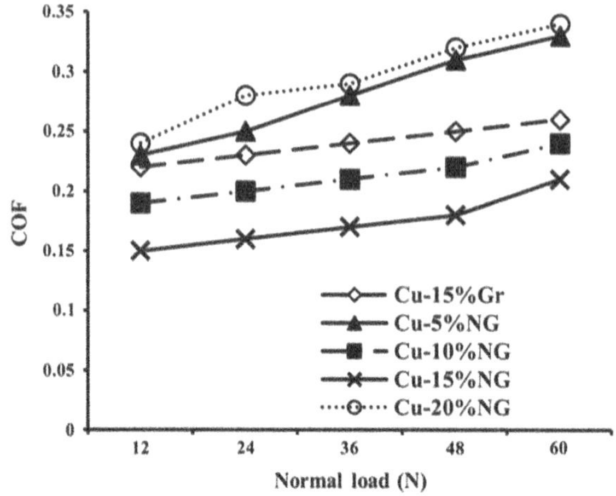

Fig. 21 Variation of coefficient of friction with normal load at sliding speed 0.77 m/s [123]

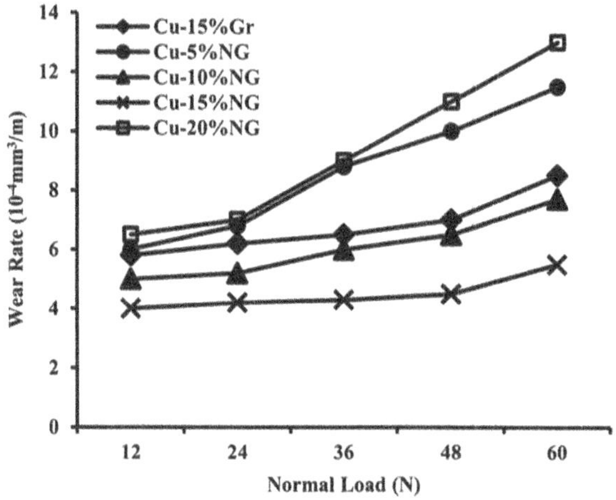

Fig. 22 Variation of wear rate of the composites with normal load at sliding speed 0.77 m/s [123]

nanographite (20 vol%), agglomeration of particles occurs and it leads to incomplete spreading of graphite at the contact zone, and hence, increases both COF and the wear rate.

Some researchers have studied the influence of different solid lubricant particles on the tribological behavior of copper matrix composites. For instance, Chen et al. [124] have investigated the tribological properties of two different solid lubricants, graphite and h-BN, for copper composites. Copper matrix composites reinforced by graphite at weight fractions in the range of 0, 2, 5, 8, and 10 %, corresponding to the hexagonal boron nitride (h-BN) at weight fractions in the range of 10, 8, 5, 2, and 0 %, respectively. Figure 23 exhibits the variation of coefficient of friction and wear rates at different normal loads. Results showed that the lubrication effects of graphite are superior to those of h-BN. Also, composites with higher amount of graphite significantly show better wear rates and friction coefficient.

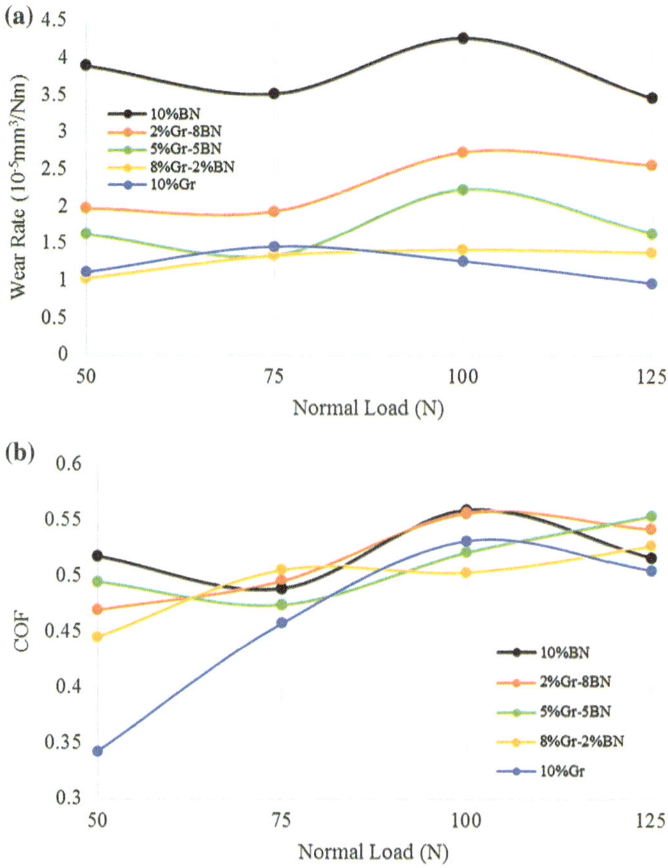

Fig. 23 Variation of **a** wear rates, and **b** friction coefficient with loads for various copper-based composites [124]

Figure 24 shows the crystalline structures of graphite and h-BN [125]. C–C bonds in graphite and B–N bonds in h-BN can be observed between neighboring interlayers of crystalline structure. Strong electrical dipoles between B and N atoms make a strong van der Waals force between interlayers of h-BN compared to graphite layer. Therefore, the interplanar spacing of h-BN is shorter than that of graphite where the distance between the neighboring interlayer of h-BN and graphite is 3.33 Å [91] and 3.35 Å [92], respectively. This shorter interplanar spacing resulted in a strong bonding in the crystalline structure of h-BN which could tend to different lubrication effects when used in copper-based composites. Weaker interlayer bonding of graphite gives the ability to easily shear along the basal plane of the crystalline structures than the h-BN. Accordingly, the wear rate of sample with 8 and 10 volume fraction of graphite is much lower than those of others due to the formation of compact and continuously tribo-films on the samples with 8 and 10 % of graphite (Fig. 25).

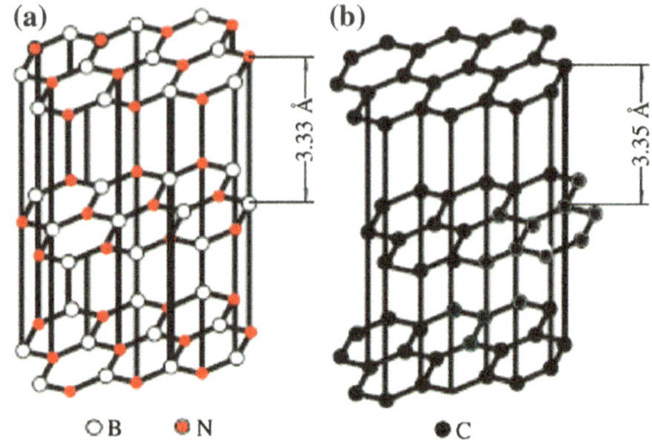

Fig. 24 Crystalline structures of **a** h-BN and **b** graphite [13]

3.3 Magnesium Matrix Composites

Magnesium alloys possess excellent properties, including low density, high specific strength and stiffness, good damping characteristic, excellent machineability and castability tends to have applications in several industries, such as automobile and aerospace. Conversely, low corrosion resistance [126] and wear resistance [127] minimize its application as widely as aluminum alloys. The influence of embedding graphite particles into magnesium matrix on the friction and wear characteristics of AZ91 magnesium alloy matrix composite was investigated by Qi [128]. Figures 26 and 27 show the effect of graphite content and normal load on the wear rate and COF of self-lubricating magnesium composites, respectively. As regards to the wear rate and COF variation, the wear rate and COF of composite enhanced in comparison with magnesium alloy under similar testing conditions. In addition, embedding more graphite tends to reduce the wear mass loss of each composite. The reason for improving the wear loss and COF is a result of the formation of continuous black lubricating film on the worn surface during sliding. This graphite film can effectively decrease the direct interaction between the composite tribo-surface and the counterpart material. Therefore, the transition from mild wear to severe wear for magnesium alloy composites will be remarkably postponed.

Zhang et al. [59] investigated the effect of graphite particle size on tribological properties of AZ91D-0.8 % Ce hybrid composites reinforced by graphite as solid lubricant and Al_2O_3 as hard ceramic particle. The variation of wear loss with load was shown in Fig. 28. It was clearly observed that the wear loss of the composites decreases by increasing the graphite particle size. Specifically, at low applied load, the composites with larger graphite size show slight improvement in the wear loss while improvement in wear loss of the composite with largest graphite particle size is significant at higher normal load. At low load, abrasive wear and oxidation wear were the dominant wear mechanisms for all the composites while at higher loads the wear mechanisms of the composites were delamination wear.

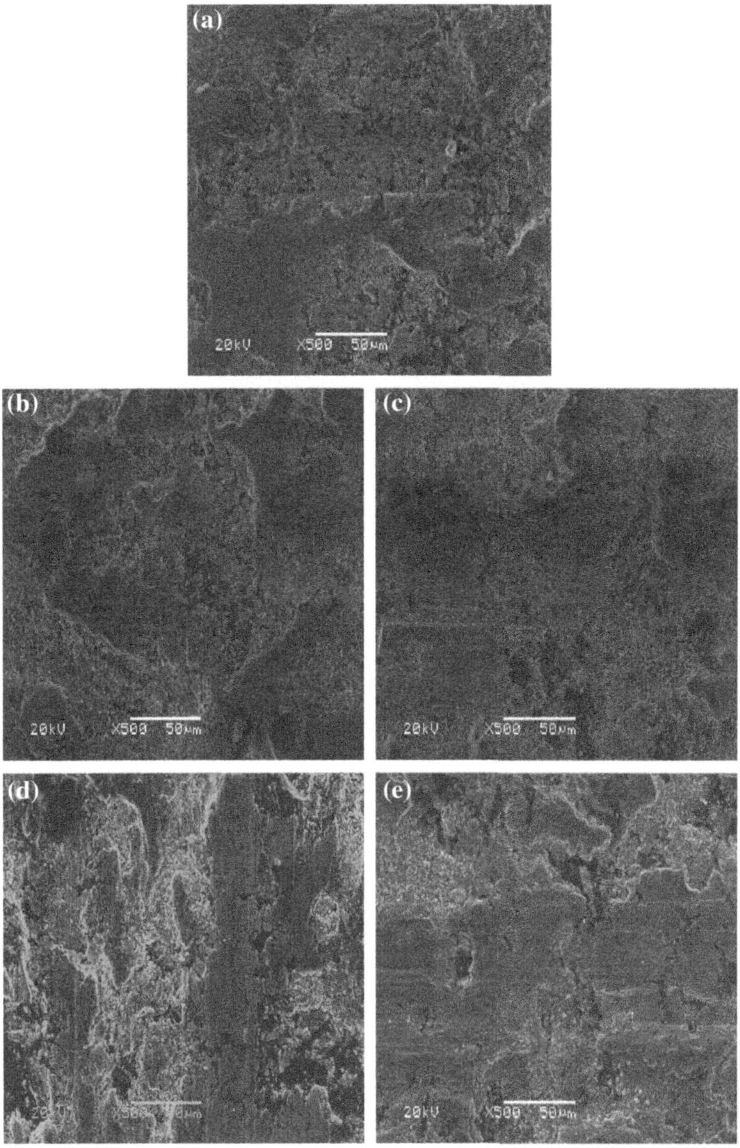

Fig. 25 SEM micrographs of the worn surfaces of the composites tested at a load of 100 N and a speed of 2.6 ms^{-1}: **a** 10 % BN; **b** 2 %Gr–8 % BN; **c** 5 % Gr–5 % BN; **d** 8 % Gr–2 % BN; and **e** 10 % Gr [124]

Mindivan et al. [129] investigated the effect of nano-solid lubricant on tribological properties of magnesium self-lubricating materials. When 0.5 wt% CNTs embedded into the magnesium matrix a significant improvements in the COF and wear rate were observed. Further, increasing the amount of CNTs tends to reduce

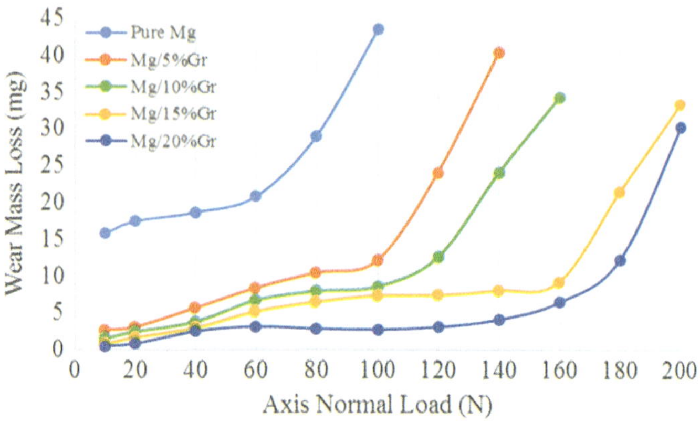

Fig. 26 Relation between wear mass loss and load [128]

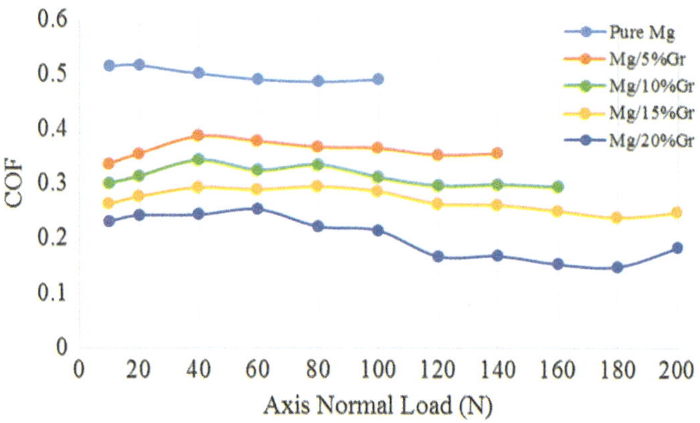

Fig. 27 Relation between friction coefficient and load [128]

the wear rate and COF as shown in Fig. 29. The maximum effective content of CNTs is 2 and 4 wt% where a reduction in the wear rate and COF are found to be the highest. Figure 30 presents the SEM and 2D-profile images of the wear tracks formed on the base alloy and composite containing 2 wt% CNTs. The morphology of worn surface of base alloy is rough, while the composite with 2 wt% CNTs showed a relatively smoother appearance (Fig. 30). The depth of the wear track of the composite containing 2 wt% CNT is less than that of the depth of the wear track of the base alloy. With the addition of CNTs, a notable reduction in the presence of craters and the formation of thin and adherent transfer film on the worn surface were observed (Fig. 30). Hence, the enhancement of the wear properties may come from the presence of CNTs that acts as lubricant medium reducing the coefficient of friction.

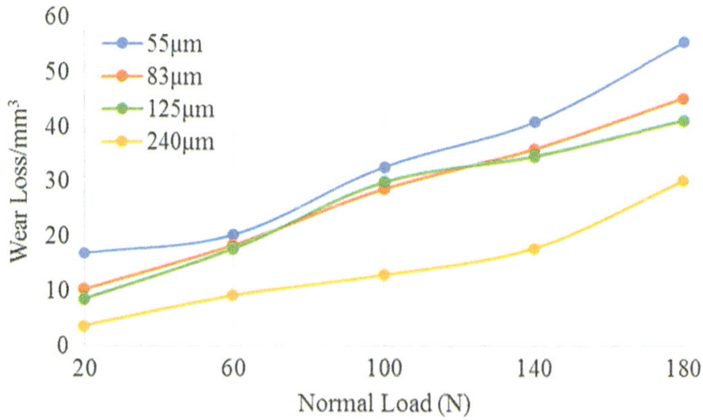

Fig. 28 Variations of wear loss with load of composites [59]

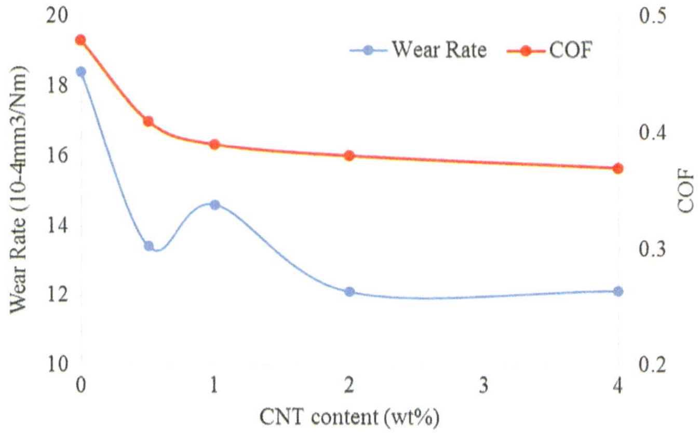

Fig. 29 Effect of CNTs on wear rate and COF of magnesium nanocomposites [129]

3.4 Nickel Matrix Composites

Excellent high-temperature performance of Nickel–graphite composites makes application of these composites for high-efficiency engine [130, 131]. Attempts have been made to study the effect of materials and testing parameters on the tribological behavior of nickel–graphite composites. Li et al. [65] studied the effect of temperature, load, and speed on the wear behavior of nickel/graphite composite. The variation of COF and wear rates of nickel/graphite composite as a function of the graphite volume fraction, load, and speed are shown in Fig. 31. Adding graphite particles to the nickel matrix show a significant improvement in the friction and wear properties when compared to pure nickel. Graphite particles with 6–12 wt% were found to be the optimum range in order to achieve the lowest COF and wear

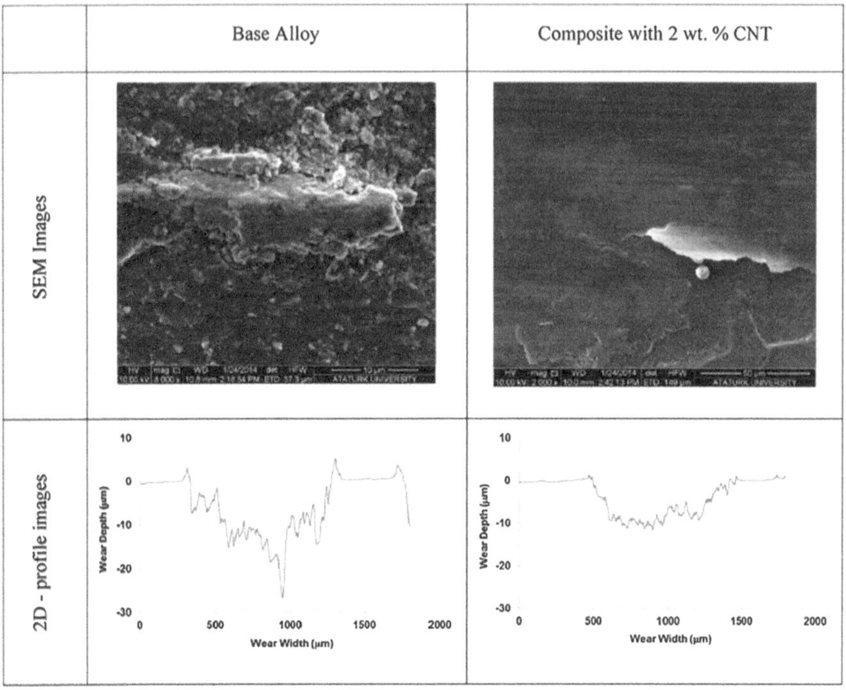

Fig. 30 SEM and 2D-profile images of the wear tracks formed on the base alloy and composite containing 2 wt% CNT [129]

rate. As shown in Fig. 31, increasing the load and sliding speed leads to reduction in friction coefficients while the wear rates increase with increasing temperature and sliding speed. In addition, the wear rate is increased by increasing the normal load up to 150 N and then decreased by increasing load.

Further, Li et al. [66] studied the tribological properties of Ni–Cr–W–Fe–C self-lubricating composite reinforced by graphite and molybdenum disulfide as solid lubricants at different temperatures. It was found that chromium sulfide and tungsten carbide were formed in the composite when adding molybdenum disulfide and graphite, which were responsible for low friction and high wear resistance at elevated temperatures, respectively. Figure 32 shows the variation of COF and wear rate for different composites at different temperatures. Among the composites, nickel composites reinforced by graphite and MoS_2 exhibited better tribological properties for a wide range of temperature due to the synergistic lubricating effect of graphite and molybdenum disulfide. The graphite played the key role of lubrication at room temperature, while sulfides were responsible for low friction at high temperatures.

Chen et al. [132] and Scharf et al. [133] investigated the effect of CNTs on the tribological behavior of self-lubricating nickel composites. The results showed a reduction in COF by embedding CNTs nanoparticles. In addition, Scharf et al.

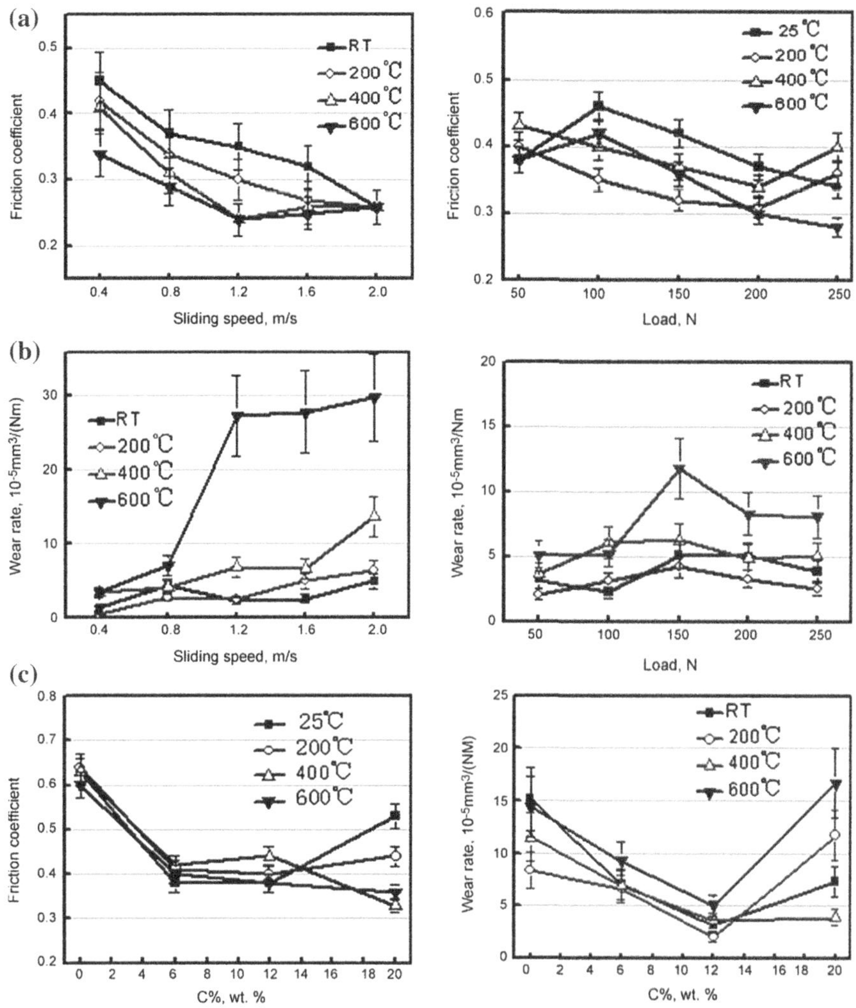

Fig. 31 Variation of COF and wear rates for different **a** graphite content **b** load and **c** speed [65]

[133] discovered that adding CNTs to the nickel matrix is more efficient than adding graphite microparticles to the nickel matrix due to strong mechanical properties of CNTs. CNTs comprise of concentric cylindrical layers or shells of graphite-like sp^2-bonded, where the intershell interactions are predominately van der Waals. These cylinders can easily slide or rotate on each other, leading to lower friction coefficient. Figure 33 shows the variation of friction coefficients and wear rate with load at different concentrations of CNTs. The results revealed that the COF decreases with increasing the load while the wear rate increases with increasing the load. Also, the optimum concentration of CNTs in nickel matrix is 1.1 g/L where the COF and wear rate possess minimum value. Figure 34 compares

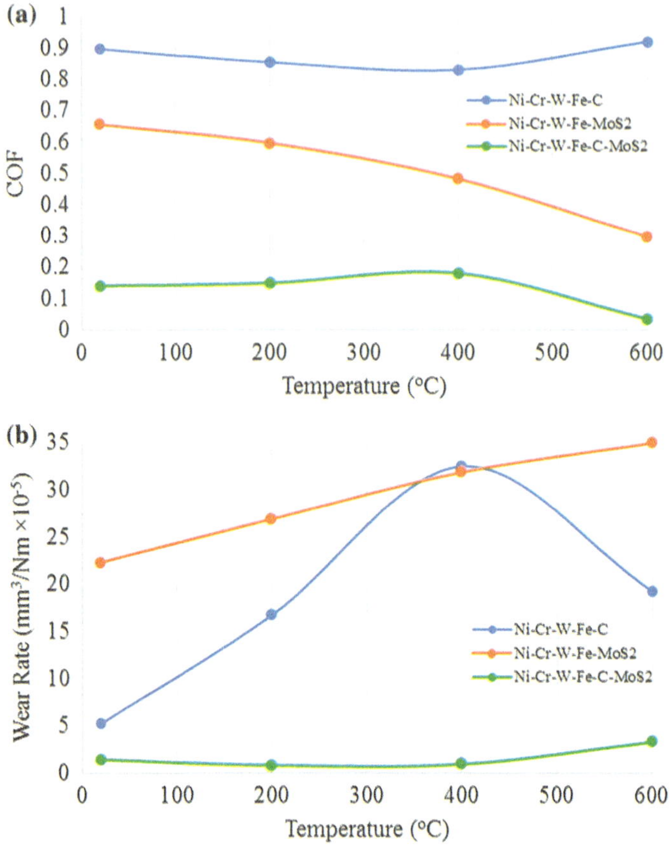

Fig. 32 Variation of **a** COF and **b** wear rate in the composites reinforced by graphite and MoS_2 [66]

the worn surface of Ni, Ni/graphite and Ni/CNTs composites. Deep grooves and a considerable extent of peeling on the worn surface of the Ni were observed. By comparing the worn surface of nickel composites reinforced by graphite and CNTs, it is revealed that there are less cracks and finer scratches on the surface of nickel/CNTs when compared to nickel/graphite, resulting in lesser amount of material loss by embedding CNTs in nickel matrix.

4 Summary

Due to concern on the environmental pollution in recent decade, there is a demand for materials that need to be environmental friendly and help to limit global warming and environmental pollution. As liquid- or grease-based lubricants are

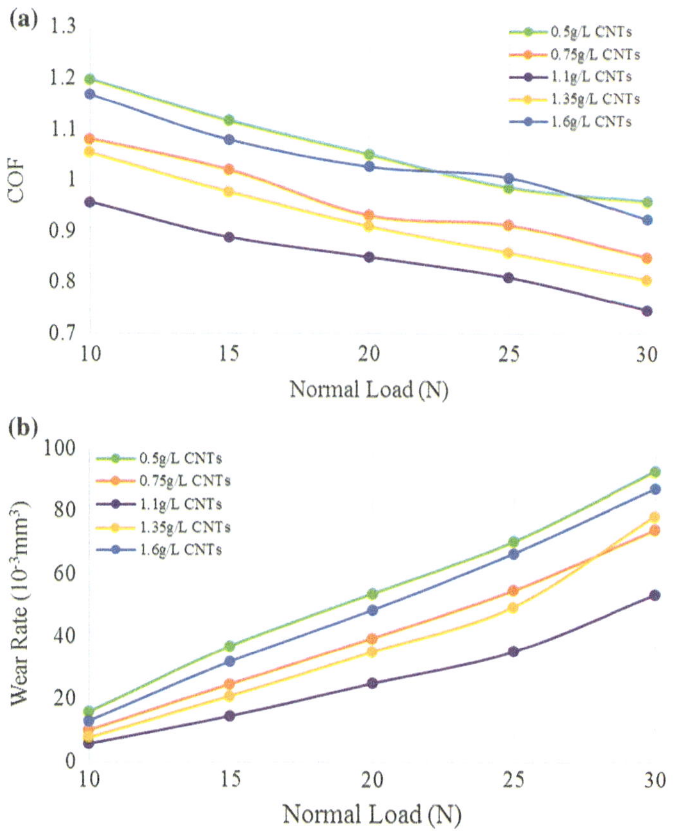

Fig. 33 Effect of CNTs concentration on **a** COF and **b** wear rate of Ni/CNTs composites [133]

used to minimize friction and wear, these lubricants can cause environmental damage. Therefore, solid lubricants, such as molybdenum disulfide, carbonous allotropes, hexagonal boron nitride, and boric acid are good substitutes for conventional lubricants. The recent development to have a continuous source of solid lubricant to the contact surface during sliding is to embed solid lubricants as reinforcement into the matrix to manufacture self-lubricating composites. The reason for the development of self-lubricating composites is the formation of lubricant tribofilm of solid lubricant during sliding that can provide adequate lubrication between contacting surfaces. Self-lubricating composites have many advantages and it is clear that research and development must continue to create the best possible innovative materials. The main function of self-lubricating composites is their ability to reduce friction and wear. This chapter points out the important parameters that affect friction and wear behavior of prominent SLMMCs including aluminum, copper, magnesium, and nickel matrix composites. Specifically, the influence of reinforcement type, reinforcement size, volume fraction of reinforcement, normal load, sliding speed, and temperature on friction and wear behavior of

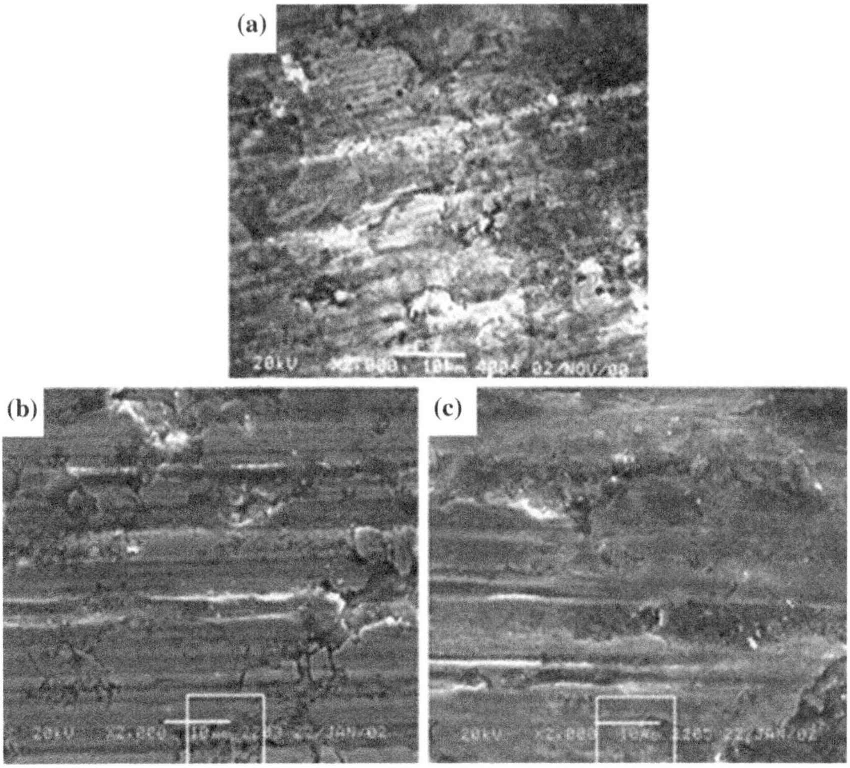

Fig. 34 Wear morphology of: **a** Ni; **b** Ni/graphite; **c** Ni/CNTs coatings [133]

SLMMCs have been discussed. Self-lubricating metal matrix composites usually exhibit superior tribological properties, such as coefficient of friction and wear rate than unreinforced alloys, and this is primarily due to the presence of solid lubricants as reinforcement in the matrix. During sliding, these solid lubricants expose to surface and emanate to the contact surface to make a lubricant film. This film acts as lubricant to decrease the metal-to-metal contact and tends to enhance tribological behavior.

In general, increasing the volume fraction of reinforcements can improve wear resistance of self-lubricating composites and decrease the friction coefficient due to more uniform lubricant film at higher volume fraction of solid lubricants. However, an increase in the volume fraction of micron-sized reinforcements over 20 vol% has no significant improvement on the wear rate. Studies have shown a reduction in the wear rate of the SLMMCs with an increase in the particle size of the reinforcements. The greatest disadvantage of metal/solid lubricant self-lubricating composites is the poor mechanical properties. Thus, embedding nanosized solid lubricants, such as carbon nanotubes (CNTs) and nanographite or graphene into metal matrix can improve mechanical, electrical, and tribological properties.

Normal load is an important mechanical factor that affects the tribological behavior of the self-lubricating composites. The results have shown a reduction in COF with increasing the normal load because of more amount of solid lubricants expose onto the contact surface. On the other hand, wear rate was increased by increasing the applied load. Similar to the normal load, sliding speed also has effect on tribological properties. An increase in the wear rate and a decrease in the friction coefficient of the SLMMCs occurred with increasing the sliding speed during the wear test.

References

1. Sasaki S (2010) Environmentally friendly tribology (eco-tribology). J Mech Sci Technol 24:67–71
2. Stojilković M, Golubović D, Ješić D (2014) Ecotribology aspects in the lubricants application
3. Taylor C (1998) Automobile engine tribology—design considerations for efficiency and durability. Wear 221:1–8
4. Tzanakis I, Hadfield M, Thomas B, Noya S, Henshaw I, Austen S (2012) Future perspectives on sustainable tribology. Renew Sustain Energy Rev 16:4126–4140
5. Bartz WJ (2006) Ecotribology: environmentally acceptable tribological practices. Tribol Int 39:728–733
6. Donnet C, Erdemir A (2004) Historical developments and new trends in tribological and solid lubricant coatings. Surf Coat Technol 180:76–84
7. Donnet C, Erdemir A (2004) Solid lubricant coatings: recent developments and future trends. Tribol Lett 17:389–397
8. Dorri Moghadam A, Omrani E, Menezes PL, Rohatgi PK (2015) Mechanical and tribological properties of self-lubricating metal matrix nanocomposites reinforced by carbon nanotubes (CNTs) and graphene–a review. Compos Part B: Eng 77:402–420
9. Menezes PL, Reeves CJ, Lovell MR (2013) Fundamentals of lubrication. In: Menezes PL, Ingole SP, Nosonovsky M, Kailas SV, Lovell MR (eds) Tribology for scientists and engineers. Springer, New York, pp 295–340
10. Menezes PL, Nosonovsky M, Kailas SV, Lovell MR (2013) Friction and wear. In: Menezes PL, Ingole SP, Nosonovsky M, Kailas SV, Lovell MR (eds) Tribology for scientists and engineers. Springer, New York, pp 43–91
11. Ludema KC (2010) Friction, wear, lubrication: a textbook in tribology. CRC press
12. Stachowiak GP, Stachowiak GW, Podsiadlo P (2008) Automated classification of wear particles based on their surface texture and shape features. Tribol Int 41:34–43
13. Moghadam AD, Schultz BF, Ferguson J, Omrani E, Rohatgi PK, Gupta N (2014) Functional metal matrix composites: self-lubricating, self-healing, and nanocomposites-an outlook. JOM 66:872–881
14. Rohatgi PK, Afsaneh DM, Schultz BF, Ferguson J (2013) Synthesis and properties of metal matrix nanocomposites (MMNCS), syntactic foams, self lubricating and self-healing metals. In: PRICM: 8 Pacific Rim international congress on advanced materials and processing. Wiley, New York, 1515–1524
15. Shankara A, Menezes PL, Simha K, Kailas SV (2008) Study of solid lubrication with MoS_2 coating in the presence of additives using reciprocating ball-on-flat scratch tester. Sadhana 33:207–220

16. Lovell MR, Kabir M, Menezes PL, Higgs CF (2010) Influence of boric acid additive size on green lubricant performance. Philos Trans R Soc Lond A: Math Phys Eng Sci 368:4851–4868
17. Reeves CJ, Menezes PL, Lovell MR, Jen T-C (2013) The size effect of boron nitride particles on the tribological performance of biolubricants for energy conservation and sustainability. Tribol Lett 51:437–452
18. Reeves CJ, Menezes PL, Lovell MR, Jen T-C (2015) The influence of surface roughness and particulate size on the tribological performance of bio-based multi-functional hybrid lubricants. Tribol Int 88:40–55
19. Menezes PL, Wornyoh EY, Lovell MR (2011) On the size effect of boron nitride particles on tribological performance of canola oil. In: ASME/STLE 2011 international joint tribology conference, Los Angeles, California, USA
20. Reeves CJ, Menezes PL, Jen T-C, Lovell MR (2013) The effect of surface roughness on the tribological performance of environmentally friendly bio-based lubricants with varying particle size. In: STLE annual meeting and exhibition
21. Reeves CJ, Menezes PL, Lovell MR, Jen T-C, Garvey SL, Dietz ML (2013) The tribological performance of bio-based room temperature ionic liquid lubricants: a possible next step in biolubricant technology. In: 5th world tribology congress
22. Rohatgi PK, Tabandeh-Khorshid M, Omrani E, Lovell MR, Menezes PL (2013) Tribology of metal matrix composites. In: Menezes PL, Ingole SP, Nosonovsky M, Kailas SV, Lovell MR (eds) Tribology for scientists and engineers. Springer, New York, pp 233–268
23. Reeves CJ, Menezes PL, Lovell MR, Jen T-C (2013) Tribology of solid lubricants. In: Menezes PL, Ingole SP, Nosonovsky M, Kailas SV, Lovell MR (eds) Tribology for scientists and engineers. Springer, New York, pp 447–494
24. Rohatgi PK, Menezes PL, Lovell MR, Kailas ASV (2014) Addition of solid lubricants to metal matrices and liquid lubricants to improve tribological performance. In: ASIATRIB—2014
25. Menezes PL, Lovell MR, Kabir M, Higgs CF III, Rohatgi PK (2012) Green lubricants: role of additive size. Green tribology. Springer, New York, pp 265–286
26. Menezes PL, Rohatgi PK, Lovell MR (2012) Self-lubricating behavior of graphite reinforced metal matrix composites. Green tribology. Springer, New York, pp 445–480
27. Menezes PL, Reeves CJ, Rohatgi PK, Lovell MR (2013) Self-lubricating behavior of graphite-reinforced composites. In: Menezes PL, Ingole SP, Nosonovsky M, Kailas SV, Lovell MR (eds) Tribology for scientists and engineers. Springer, New York, pp 341–389
28. Mannekote JK, Menezes PL, Kailas SV, Sathwik RC (2013) Tribology of green lubricants. In: Menezes PL, Ingole SP, Nosonovsky M, Kailas SV, Lovell MR (eds) Tribology for scientists and engineers. Springer, New York, pp 495–521
29. Reeves CJ, Menezes PL, Jen T-C, Lovell MR (2012) Evaluating the tribological performance of green liquid lubricants and powder additive based green liquid lubricants. In: Proceedings of 2012 STLE annual meeting and exhibition, STLE 2012
30. Patnaik S, Swain P, Mallik P, Sahoo S (2014) Wear characteristics of aluminium-graphite composites produced by stir casting technique. J Mater Metall Eng 4:13–20
31. Dwivedi SK, Patel S (2014) Evaluation of hardness of aluminium/graphite particulate composite fabricated by stir casting route. Evaluation 3:26–28
32. Shanmughasundaram P, Subramanian R (2013) Wear behaviour of eutectic Al-Si alloy-graphite composites fabricated by combined modified two-stage stir casting and squeeze casting methods. Adv Mater Sci Eng
33. Wu LL, Yang WJ, Xu JR, Yao GC (2013) Wear resistance of graphite/aluminium composites that prepared by stirring casting. Adv Mater Res: Trans Tech Publ 2013:333–338
34. Baghani A, Davami P, Varahram N, Shabani MO (2014) Investigation on the effect of mold constraints and cooling rate on residual stress during the sand-casting process of 1086 steel by employing a thermomechanical model. Metall Mater Trans B 45:1157–1169
35. Baghani A, Bahmani A, Davami P, Varahram N, Shabani MO (2013) Application of computational fluid dynamics to study the effects of sprue base geometry on the surface and

internal turbulence in gravity casting. Proc Inst Mech Eng Part L: J Mater Des Appl 2013:1464420713500182
36. Etter T, Kuebler J, Frey T, Schulz P, Löffler J, Uggowitzer P (2004) Strength and fracture toughness of interpenetrating graphite/aluminium composites produced by the indirect squeeze casting process. Mater Sci Eng A 386:61–67
37. Zeren A (2015) Effect of the graphite content on the tribological properties of hybrid Al/SiC/Gr composites processed by powder metallurgy. Ind Lubr Tribol 67:262–268
38. Ravindran P, Manisekar K, Narayanasamy R, Narayanasamy P (2013) Tribological behaviour of powder metallurgy-processed aluminium hybrid composites with the addition of graphite solid lubricant. Ceram Int 39:1169–1182
39. Chen J, Huang I (2013) Thermal properties of aluminum–graphite composites by powder metallurgy. Compos B Eng 44:698–703
40. Mahdavi S, Akhlaghi F (2011) Effect of the graphite content on the tribological behavior of Al/Gr and Al/30SiC/Gr composites processed by in situ powder metallurgy (IPM) method. Tribol Lett 44:1–12
41. Akhlaghi F, Zare-Bidaki A (2009) Influence of graphite content on the dry sliding and oil impregnated sliding wear behavior of Al 2024–graphite composites produced by in situ powder metallurgy method. Wear 266:37–45
42. Akhlaghi F, Mahdavi S (2011) Effect of the SiC content on the tribological properties of hybrid Al/Gr/SiC composites processed by in situ powder metallurgy (IPM) method. Adv Mater Res 264–265:1878–1886
43. Hashemabad SG, Ando T (2015) Ignition characteristics of hybrid Al–Ni–Fe 2 O 3 and Al–Ni–CuO reactive composites fabricated by ultrasonic powder consolidation. Combust Flame 162:1144–1152
44. Omrani E, Dorri Moghadam A, Menezes PL, Rohatgi PK (2015) Influences of graphite reinforcement on the tribological properties of self-lubricating aluminum matrix composites for green tribology, sustainability, and energy efficiency—a review. Int J Adv Manuf Technol. doi:10.1007/s00170-015-7528-x
45. Rohatgi P, Ray S, Liu Y (1992) Tribological properties of metal matrix-graphite particle composites. Int Mater Rev 37:129–152
46. El-Hajjar RF, Shams SS, Kehrl DJ (2013) Closed form solutions for predicting the elastic behavior of quasi-isotropic triaxially braided composites. Compos Struct 101:1–8
47. Shams SS, El-Hajjar RF (2013) Effects of scratch damage on progressive failure of laminated carbon fiber/epoxy composites. Int J Mech Sci 67:70–77
48. Qamhia II, Shams SS, El-Hajjar RF (2014) Quasi-isotropic triaxially braided cellulose reinforced composites. Mech Adv Mater Struct
49. Vernet N, Ruiz E, Advani S, Alms J, Aubert M, Barburski M et al (2014) Experimental determination of the permeability of engineering textiles: benchmark II. Compos A Appl Sci Manuf 61:172–184
50. Barari B, Pillai K (2014) Calibration of one-dimensional flow setup used for estimating fabric permeability using three different reference media. Polym Compos
51. Etaati A, Pather S, Fang Z, Wang H (2014) The study of fibre/matrix bond strength in short hemp polypropylene composites from dynamic mechanical analysis. Compos B Eng 62: 19–28
52. Etaati A, Mehdizadeh SA, Wang H, Pather S (2013) Vibration damping characteristics of short hemp fibre thermoplastic composites. J Reinf Plast Compos 2013:0731684413512228
53. Rad BA, Alizadeh P (2009) Pressureless sintering and mechanical properties of SiO_2–Al_2O_3–MgO–K_2O-TiO_2–F (CaO–Na_2O) machinable glass–ceramics. Ceram Int 35:2775–2780
54. Omrani E, Barari B, Dorri Moghadam A, Rohatgi PK, Pillai KM (2015) Mechanical and tribological properties of self-lubricating bio-based carbon-fabric epoxy composites made using liquid composite molding. Tribol Int 92:222–232. doi:10.1016/j.triboint.2015.06.007
55. Xiao Y, Shi X, Zhai W, Yang K, Yao J (2015) Effect of temperature on tribological properties and wear mechanisms of NiAl matrix self-lubricating composites containing graphene nanoplatelets. Tribol Trans 58:729–735

56. Prabhu TR, Varma V, Vedantam S (2014) Effect of reinforcement type, size, and volume fraction on the tribological behavior of Fe matrix composites at high sliding speed conditions. Wear 309:247–255
57. Prabhu TR, Varma V, Vedantam S (2014) Effect of SiC volume fraction and size on dry sliding wear of Fe/SiC/graphite hybrid composites for high sliding speed applications. Wear 309:1–10
58. Karamış M, Cerit AA, Selçuk B, Nair F (2012) The effects of different ceramics size and volume fraction on wear behavior of Al matrix composites (for automobile cam material). Wear 289:73–81
59. Zhang M-J, Liu Y-B, Yang X-H, Jian A, Luo K-S (2008) Effect of graphite particle size on wear property of graphite and Al_2O_3 reinforced AZ91D-0.8 % Ce composites. Trans Nonferrous Metals Soc China 18:s273–s277
60. Bolvardi H, Khorsand H, Movahed P, Etaati A (2010) Effects of load and sliding speed on tribological behavior of plasma sprayed bronze-alumina coatings. Defect Diff Forum: Trans Tech Publ 1122–1126
61. Prabhu TR (2015) Effects of solid lubricants, load, and sliding speed on the tribological behavior of silica reinforced composites using design of experiments. Mater Des 77:149–160
62. Khun N, Frankel G, Sumption M (2014) Effects of normal load, sliding speed, and surface roughness on tribological properties of niobium under dry and wet conditions. Tribol Trans 57:944–954
63. Xu Z, Shi X, Zhang Q, Zhai W, Li X, Yao J et al (2014) Effect of sliding speed and applied load on dry sliding tribological performance of TiAl matrix self-lubricating composites. Tribol Lett 55:393–404
64. Baradeswaran A, Elayaperumal A (2011) Wear characteresitic of Al6061 reinforced with graphite under different loads and speeds. Adv Mater Res 287–290:998–1002
65. Li J, Xiong D (2009) Tribological behavior of graphite-containing nickel-based composite as function of temperature, load and counterface. Wear 266:360–367
66. Li JL, Xiong DS (2008) Tribological properties of nickel-based self-lubricating composite at elevated temperature and counterface material selection. Wear 265:533–539
67. Li J-L, Xiong D-S, Wan Y (2007) Tribological properties of nickel-graphite composite against different counterfaces at elevated temperatures. Trans Nonferrous Metals Soc China 17:s99–s104
68. Skeldon P, Wang H, Thompson G (1997) Formation and characterization of self-lubricating MoS_2 precursor films on anodized aluminium. Wear 206:187–196
69. Rawal SP (2001) Metal-matrix composites for space applications. JOM 53:14–17
70. Reboul M, Baroux B (2011) Metallurgical aspects of corrosion resistance of aluminium alloys. Mater Corros 62:215–233
71. Molina J-M, Rhême M, Carron J, Weber L (2008) Thermal conductivity of aluminum matrix composites reinforced with mixtures of diamond and SiC particles. Scripta Mater 58: 393–396
72. Recoules V, Renaudin P, Clérouin J, Noiret P, Zérah G (2002) Electrical conductivity of hot expanded aluminum: experimental measurements and ab initio calculations. Phys Rev E 66:056412
73. Li G, Ma Y, He X, Li W, Li P (2012) Damping capacity of high strength-damping aluminum alloys prepared by rapid solidification and powder metallurgy process. Trans Nonferrous Metals Soc China 22:1112–1117
74. Rohatgi P, Menezes P, Mazzei T, Lovell M (2011) Tribological behavior of aluminum micro-and nano-composites. Int J Aerosp Innov 3:153–162
75. Rohatgi PK, Menezes PL, Mazzei T, Lovell MR (2011) Tribological performance of aluminum micro and nano composites. In: ASME/STLE 2011 international joint tribology conference: American society of mechanical engineers, pp 257–259
76. Erfanian-Naziftoosi HR, Haghdadi N, Kiani-Rashid AR (2012) The effect of isothermal heat treatment time on the microstructure and properties of 2.11 % Al austempered ductile iron. J Mater Eng Perform 21:1785–1792

77. Haghdadi N, Bazaz B, Erfanian-Naziftoosi H, Kiani-Rashid A (2012) Microstructural and mechanical characteristics of Al-alloyed ductile iron upon casting and annealing. Int J Miner Metall Mater 19:812–820
78. Bonollo F, Carturan I, Capitò G, Molina R (2013) Life cycle assessment in the automotive industry: comparison between aluminium and cast iron cylinder blocks. Metall Sci Technol 24
79. Kumar GV, Rao C, Selvaraj N (2011) Mechanical and tribological behavior of particulate reinforced aluminum metal matrix composites–a review. J Miner Mater Charact Eng 10:59
80. Kathiresan M, Sornakumar T (2010) Friction and wear studies of die cast aluminum alloy-aluminum oxide-reinforced composites. Ind Lubr Tribol 62:361–371
81. Rohatgi P, Ray S, Liu Y (1992) Tribological properties of metal matrix-graphite particle composites. Int Mater Rev 37:129–152
82. Ames W, Alpas A (1995) Wear mechanisms in hybrid composites of graphite-20 Pct SiC in A356 aluminum alloy (Al-7 Pct Si-0.3 Pct Mg). Metall Mater Trans A 26:85–98
83. Roy M, Venkataraman B, Bhanuprasad V, Mahajan Y, Sundararajan G (1992) The effect of participate reinforcement on the sliding wear behavior of aluminum matrix composites. Metall Trans A 23:2833–2847
84. Alpas A, Zhang J (1994) Effect of microstructure (particulate size and volume fraction) and counterface material on the sliding wear resistance of particulate-reinforced aluminum matrix composites. Metall Mater Trans A 25:969–983
85. Gül H, Kılıç F, Uysal M, Aslan S, Alp A, Akbulut H (2012) Effect of particle concentration on the structure and tribological properties of submicron particle SiC reinforced Ni metal matrix composite (MMC) coatings produced by electrodeposition. Appl Surf Sci 258:4260–4267
86. Van Acker K, Vanhoyweghen D, Persoons R, Vangrunderbeek J (2005) Influence of tungsten carbide particle size and distribution on the wear resistance of laser clad WC/Ni coatings. Wear 258:194–202
87. Ghasemi-kahrizsangi A, Kashani-Bozorg S, Moshref-Javadi M (2015) Effect of friction stir processing on the tribological performance of steel/Al_2O_3 nanocomposites. Surf Coat Technol
88. Akhlaghi F, Pelaseyyed SA (2004) Characterization of aluminum/graphite particulate composites synthesized using a novel method termed “in-situ powder metallurgy”. Mater Sci Eng A 385:258–266
89. Nayeb-Hashemi H, Seyyedi J (1989) Study of the interface and its effect on mechanical properties of continuous graphite fiber-reinforced 201 aluminum. Metall Trans A 20:727–739
90. Tokisue H, Abbaschian G (1978) Friction and wear properties of aluminum-particulate graphite composites. Mater Sci Eng 34:75–78
91. Ravindran P, Manisekar K, Kumar SV, Rathika P (2013) Investigation of microstructure and mechanical properties of aluminum hybrid nano-composites with the additions of solid lubricant. Mater Des 51:448–456
92. Shanmughasundaram P, Subramanian R (2013) Wear behaviour of eutectic Al-Si alloy-graphite composites fabricated by combined modified two-stage stir casting and squeeze casting methods. Adv Mater Sci Eng 1–8
93. Suresha S, Sridhara BK (2010) Wear characteristics of hybrid aluminium matrix composites reinforced with graphite and silicon carbide particulates. Compos Sci Technol 70:1652–1659
94. Srivastava S, Mohan S, Srivastava Y, Shukla AJ (2012) Study of the wear and friction behavior of immiscible as cast-Al-Sn/graphite composite. Int J Mod Eng Res 2:25–42
95. Hocheng H, Yen SB, Ishihara T, Yen BK (1997) Fundamental turning characteristics of a tribology-favored graphite/aluminum alloy composite material. Compos A Appl Sci Manuf 28:883–890
96. Baradeswaran A, Perumal E (2014) Wear and mechanical characteristics of Al 7075/graphite composites. Compos B 56:472–476
97. Baradeswaran A, Perumal AE (2014) Study on mechanical and wear properties of Al 7075/Al_2O_3/graphite hybrid composites. Compos B 56:464–471

98. Basavarajappa S, Chandramohan G, Mukund K, Ashwin M, Prabu M (2006) Dry sliding wear behavior of Al 2219/SiCp-Gr hybrid metal matrix composites. J Mater Eng Perform 15:668–674
99. Prasad BK, Das S (1991) The significance of the matrix microstructure on the solid lubrication characteristics in aluminum alloys. Mater Sci Eng A 144:229–235
100. Radhika N, Subramanian R, Prasat SV, Anandavel B (2012) Dry sliding wear behaviour of aluminium/alumina/graphite hybrid metal matrix composites. Ind Lubr Tribol 64:359–366
101. Babić M, Stojanović B, Mitrović S, Bobić I, Miloradović N, Pantić M et al (2013) Wear properties of A356/10SiC/1Gr hybrid composites in lubricated sliding conditions. Tribol Ind 35:148–154
102. Rajaram G, Kumaran S, Rao TS (2011) Fabrication of Al–Si/graphite composites and their structure–property correlation. J Compos Mater 45:2743–2750
103. Chen Z, Chen Y, An G, Shu Q, Li D, Liu Y (2000) Microstructure and properties of in situ Al/TiB2 composite fabricated by in-melt reaction method. Metall Mater Trans A 31:1959–1964
104. Tjong SC (2007) Novel nanoparticle-reinforced metal matrix composites with enhanced mechanical properties. Adv Eng Mater 9:639–652
105. Thostenson ET, Li C, Chou T-W (2005) Nanocomposites in context. Compos Sci Technol 65:491–516
106. He F, Han Q, Jackson MJ (2008) Nanoparticulate reinforced metal matrix nanocomposites–a review. Int J Nanopart 1:301–309
107. Singh J, Narang D, Batra NK (2013) Experimental investigation of mechanical and tribological properties of Aa-SiC and Al-Gr metal matrix composite. Int J Eng Sci Technol 5:1205–1210
108. Ghasemi-Kahrizsangi A, Kashani-Bozorg SF (2012) Microstructure and mechanical properties of steel/TiC nano-composite surface layer produced by friction stir processing. Surf Coat Technol
109. Tabandeh-Khorshid M, Jenabali-Jahromi SA, Moshksar MM (2010) Mechanical properties of tri-modal Al matrix composites reinforced by nano- and submicron-sized Al_2O_3 particulates developed by wet attrition milling and hot extrusion. Mater Des 31:3880–3884
110. Shafiei-Zarghani A, Kashani-Bozorg SF, Zarei-Hanzaki A (2009) Microstructures and mechanical properties of Al/Al_2O_3 surface nano-composite layer produced by friction stir processing. Mater Sci Eng A 500(87):91
111. Jinfeng L, Longato J, Gaohui W, Shoufu T, Guoqin C (2009) Effect of graphite particle reinforcement on dry sliding wear of SiC/Gr/Al composites. Rare Metal Mater Eng 38:1894–1898
112. Bakshi S, Lahiri D, Agarwal A (2010) Carbon nanotube reinforced metal matrix composites—a review. Int Mater Rev 55:41–64
113. Kovacik J, Emmer S, Bielek J, Kelesi L (2008) Effect of composition on friction coefficient of Cu–graphite composites. Wear 256:417–421
114. S-m Zhou, X-b Zhang, Z-p Ding, Min C-y Xu, G-l Zhu W-m (2007) Fabrication and tribological properties of carbon nanotubes reinforced Al composites prepared by pressureless infiltration technique. Compos A 38:301
115. Choi HJ, Lee SM, Bae DH (2010) Wear characteristic of aluminum-based composites containing multi-walled carbon nanotubes. Wear 270:12
116. Ghazaly A, Seif B, Salem HG (2013) Mechanical and tribological properties of AA2124-graphene self lubricating nanocomposite. Light Metals 2013:411–415
117. Zamzam M (1989) Wear resistance of agglomerated and dispersed solid lubricants in aluminium. Mater Trans JIM 30:516–522
118. Dharmalingam S, Subramanian R, Vinoth KS, Anandavel B (2011) Optimization of tribological properties in aluminum hybrid metal matrix composites using gray-Taguchi method. J Mater Eng Perform 20:1457–1466
119. Moustafa S, El-Badry S, Sanad A, Kieback B (2002) Friction and wear of copper–graphite composites made with Cu-coated and uncoated graphite powders. Wear 253:699–710

120. Zhao H, Liu L, Wu Y, Hu W (2007) Investigation on wear and corrosion behavior of Cu–graphite composites prepared by electroforming. Compos Sci Technol 67:1210–1217
121. Jincheng X, Hui Y, Xiaolong L, Hua Y (2004) Effects of some factors on the tribological properties of the short carbon fiber-reinforced copper composite. Mater Des 25:489–493
122. Ramesh C, Ahmed RN, Mujeebu M, Abdullah M (2009) Development and performance analysis of novel cast copper–SiC–Gr hybrid composites. Mater Des 30:1957–1965
123. Rajkumar K, Aravindan S (2013) Tribological behavior of microwave processed copper–nanographite composites. Tribol Int 57:282
124. Chen B, Bi Q, Yang J, Xia Y, Hao J (2008) Tribological properties of solid lubricants (graphite, h-BN) for Cu-based P/M friction composites. Tribol Int 41:1145–1152
125. Petrescu M (2004) Boron nitride theoretical hardness compared to carbon polymorphs. Diam Relat Mater 13:1848–1853
126. Song G, Bowles AL, StJohn DH (2004) Corrosion resistance of aged die cast magnesium alloy AZ91D. Mater Sci Eng A 366:74–86
127. Chen H, Alpas A (2000) Sliding wear map for the magnesium alloy Mg-9Al-0.9 Zn (AZ91). Wear 246:106–116
128. Qi Q-J (2006) Evaluation of sliding wear behavior of graphite particle-containing magnesium alloy composites. Trans Nonferrous Metals Soc China 16:1135–1140
129. Mindivan H, Efe A, Kosatepe AH, Kayali ES (2014) Fabrication and characterization of carbon nanotube reinforced magnesium matrix composites. Appl Surf Sci 318:234–243
130. Etaati A, Shokuhfar A, Omrani E, Movahed P, Bolvardi H, Tavakoli H (2010) Study on homogenization time and cooling rate on microstructure and hardness of Ni-42.5 wt% Ti-3wt % Cu alloy. Defect Diff Forum: Trans Tech Publ 489–494
131. Omrani E, Shokuhfar A, Etaati A, Dorri M, Saatian A (2010) The effects of homogenization time and cooling environment on microstructure and transformation temperatures of Ni-42.5 wt% Ti-7.5 wt% Cu alloy. Defect Diff Forum: Trans Tech Publ 344–350
132. Chen X, Chen C, Xiao H, Liu H, Zhou L, Li S et al (2006) Dry friction and wear characteristics of nickel/carbon nanotube electroless composite deposits. Tribol Int 39:22–28
133. Scharf T, Neira A, Hwang J, Tiley J, Banerjee R (2009) Self-lubricating carbon nanotube reinforced nickel matrix composites. J Appl Phys 106:013508

Chapter 4
Multi-objective Optimization of Engine Parameters While Bio-lubricant–Biofuel Combination of VCR Engine Using Taguchi-Grey Approach

S. Arumugam, G. Sriram, T. Rajmohan and J. Paulo Davim

Abstract The substitution of petroleum-based synthetic lubricant with the vegetable oil-based bio-lubricant for an engine fueled with biodiesel is explored in this study. Rapeseed oil was selected as a base oil candidate for the formulation of bio-lubricant as well as biofuel. Further, this study investigated the multi-response optimization of diesel engine for an optimal parametric combination to yield better engine performance such as brake power, brake thermal, and mechanical efficiency with minimum exhaust emissions like carbon monoxide (CO), hydrocarbon (HC), and smoke and oxides of nitrogen (NO_x) when the engine is fueled with rapeseed oil biodiesel and lubricated with rapeseed oil-based bio-lubricant using Taguchi-grey relational analysis. Three factors namely, lubricant, compression ratio, and engine load were optimized using L_{18} orthogonal array. The response table, response graph, and analysis of variance (ANOVA) are used to find the optimal setting and the influence of engine parameters on the multiple performance characteristics. The optimization results have shown that an increase in value of the grey relational grade from 0.6105 to 0.85 confirms the improvement in engine characteristics when using rapeseed oil-based bio-lubricant/biodiesel combination.

S. Arumugam (✉) · G. Sriram · T. Rajmohan
Department of Mechanical Engineering, Sri Chandrasekharendra Saraswathi Viswa Mahavidyalaya, Enathur, Kanchipuram 631561, Tamil Nadu, India
e-mail: aru_amace@yahoo.co.in

G. Sriram
e-mail: drg.sriram@gmail.com

T. Rajmohan
e-mail: rajmohanscsvmv@yahoo.com

J. Paulo Davim
Department of Mechanical Engineering, University of Aveiro Campus Santiago, 3810-193 Aveiro, Portugal
e-mail: pdavim@ua.pt

J.P. Davim (ed.), *Ecotribology*, Materials Forming, Machining and Tribology,
DOI 10.1007/978-3-319-24007-7_4

1 Introduction

The rising world crude oil price, the growing environmental awareness, and the fast depleting crude oil reserves have spurred renewed research interest and advances in alternative lubricant and fuel development from renewable energy sources [1, 2]. The base oil used for the formulation of most lubricants is environmentally hostile mineral oil. But, being a product of distillation of crude oil, the mineral oil is expected to be used only as long as crude oil is available. Additionally, the disposal of mineral oil leads to pollution of ecosystem [3]. Thus the search for environmentally friendly substitutes to mineral oils as base oils in lubricants has become a frontier area of research. Vegetable oils are perceived to be alternatives to mineral oils for lubricant base oils because of certain inherent technical properties and their ability for biodegradability. Compared to mineral oils, vegetable oils in general possess high flash point, high viscosity index, high lubricity, and low evaporative loss.

Nevertheless, the principal weakness of vegetable oils for lubrication has been their tendency to oxidize at higher temperature, giving rise to gum, varnish, and sludge formation at lower temperature [4]. Many literatures revealed that technical solutions such as chemical modification and additivation have been suggested to overcome poor thermo-oxidative stability and low temperature fluidity of vegetable oils. Attempts have been made to improve the oxidative stability and low temperature fluidity by transesterification of trimethylolpropane and rapeseed oil methyl ester [5]. Efforts have been made to improve the low temperature fluidity using an additivation method of blending the vegetable oils with diluents such as poly-a-olefin, oleate, and diisodecyl adipate [6]. Alternatively, the unsaturation present in the fatty acid molecule of the vegetable oil can be used to introduce various functional groups by carrying out chemical modifications. Kim and Sharma discussed the possibilities of utilization of epoxidized products of vegetable oils in PVC formulations and bio-thermoset plastics [7]. Chen-Ching Ting and Chien-Chih Chen investigated the viscosity analysis of a mixture of epoxidized and hydrogenated soybean oils as engine lubricants [8].

Although many valuable polymeric materials and lubricants derived from chemically modified vegetable oil, the use of chemically modified vegetable oil as automotive lubricants was not found as such in India. In this row, green crankcase oil via chemical modifications of rapeseed oil such as epoxidation, hydroxylation, and esterification process was formulated by Arumugam et al. [9]. Furthermore, the issue of friction and wear characteristics of the diesel engine cylinder liner/piston ring combination under chemically modified rapeseed oil bio-lubricant using a high-frequency reciprocating tribometer is addressed in [10]. Twelve percent of higher wear was observed with chemically modified rapeseed bio-lubricant than that of commercial synthetic lubricant (SAE20W40). In order to improve the antiwear behavior of chemically modified rapeseed oil bio-lubricant, nano-copper oxide (CuO) was added [11].

From the alternative fuel point of view, it is proven that biodiesel is a promising fuel for diesel engine up to 20 % blend in connection with emission as well as performance [12]. Agarwal reported that the use of biodiesel contributes to a large reduction in engine wear and although vegetable oil creates various long-term problems in engine components [13]. Hence, the examination of the impact of vegetable oil-based bio-lubricant in a diesel engine fueled with biodiesel is still to be understood. Also, the available synthetic lubricant is dedicated to petroleum fuels. As far as the internal combustion engines are concerned, the thermal efficiency and emission are the important parameters for which the other design and operating parameters have to be optimized. The most common optimization techniques used for multi-objective problems are response surface method [14], grey relational analysis [15], nonlinear regression analysis [16], genetic algorithm [17], and Taguchi method [18].

The review of literature clearly indicated that the researchers have put sincere attempt to find out the suitable alternative to petroleum-based fuel and lubricants without going through any engine modification. Also the information about the use of bio-lubricant on engine performance and emissions is scarce. Therefore, it is of prime importance to investigate the effect of bio-lubricant on diesel engine's performance and emissions when an engine is fueled with B20 rapeseed oil biodiesel (20 % by vol. rapeseed oil methyl ester and 80 % by vol. diesel). Furthermore to investigate whether the formulated bio-lubricant is in comparable performance with synthetic lubricant or not, the objective of the present study is to determine the optimum compression ratio, lubricant type, and engine load that would result in a better engine performance along with minimum emission characteristics using Taguchi-based grey approach.

2 Materials and Methods

2.1 Formulation of Rapeseed Oil Bio-lubricant

Rapeseed oil is chemically modified via epoxidation, hydroxylation, and esterification process. The detailed procedure for the chemical modification process is adopted from our earlier study [19]. The nano-copper oxide (CuO) ~ 40–70 nm of 0.5 % w/w was dispersed in CMRO as an antiwear additive using ultrasonic sonicator. Commercially available nanoparticles supplied by M/S US Research Nanomaterial Inc., USA were used in this study. Ethylene glycol of 0.1 wt% as dispersant has been added into the mixture to ensure the homogeneity of nanoparticles. The nanoparticles were ultrasonically vibrated using ultrasonic sonicator (Model: PR-1000, 25 kHz; Make: OSCAR®) for 2 h. Figure 1a, b shows the photographic view of chemical modification setup and Table 1 shows the properties of synthetic lubricant and chemically modified rapeseed oil bio-lubricant. The scheme of chemical modification is represented as follows (Scheme 1).

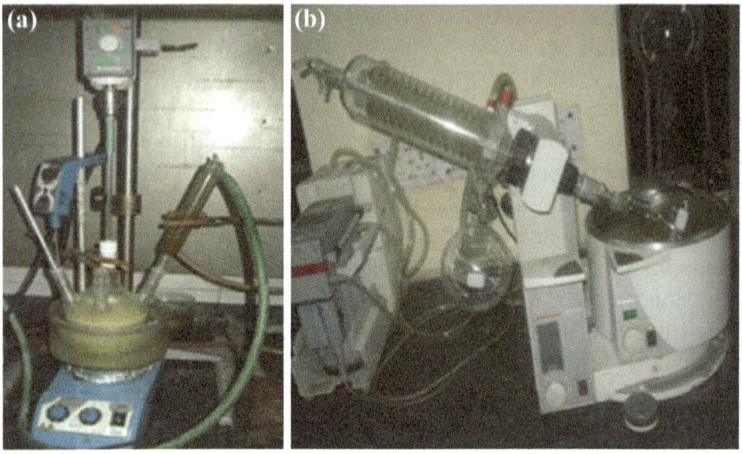

Fig. 1 **a**, **b** Photographic view of chemical modification setup

Table 1 Properties of raw and CMRO

Properties	Standard	Raw rapeseed oil	CMRO with nano CuO	SAE20W40
Viscosity @100 ° C (cSt)	ASTM D445	8	15.4	15.2
Pour point (°C)	ASTM D97	−11	−15	−21
Flash point (°C)	ASTM D92	320	242	250
Viscosity index	ASTM D2270	220	179	133
Oxirane content (%)	AOCS cd 9-57	–	5.81	5.76
Biodegradability (%)	CEC-L-33-A93	>95	>95	>95
Rotary bomb oxidation time (min)	ASTM D2272	16	35	38
Iodine value	AOCS cd 1-25	120	5.1	–

2.2 Experimental Design and Methodology

Three input parameters namely lubricant type (*A*), compression ratio (*B*), and engine load (*C*) were considered to be the main design factors for maximizing the engine performance and minimizing the exhaust emissions of a variable compression ratio (VCR) engine. The control parameters and their levels selected for the present investigation are given in Table 2. Altogether, seven (response) output parameters were analyzed, namely performance characteristics, i.e., Brake power (BP), brake thermal efficiency (BTE), and mechanical efficiency (ME); and exhaust emission characteristics such as CO, HC, NO_X, and smoke. Taguchi's parameter design was adopted to understand the effect of different input parameters on output

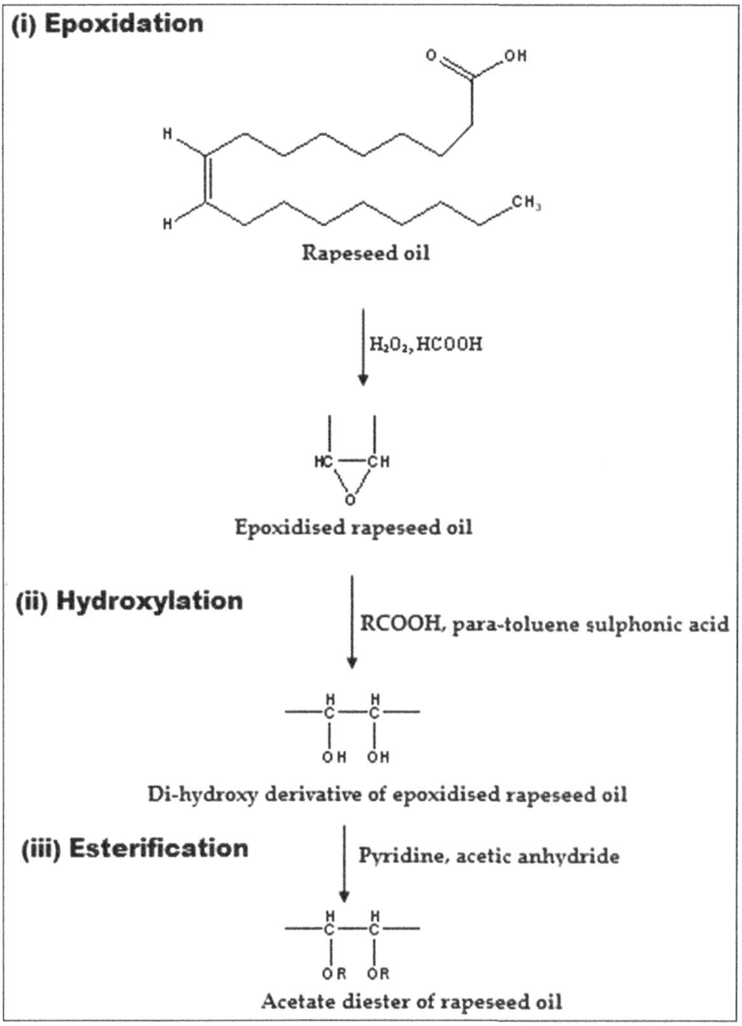

Scheme 1 Chemical modification of rapeseed oil

responses. Since multiple performance characteristics with conflicting goals were present, Taguchi-grey method was adopted to generate a single response from multiple performance characteristics.

An engine performance and emission test were executed on a computerized four-stroke, single-cylinder, water-cooled, direct injection, VCR diesel engine coupled to an eddy current dynamometer for loading purpose. An online data acquisition system is used to connect the probes through an analog-to-digital converter and the data are fed to the computer. A specialized labview-based engine analysis software "Engine Soft LV" has been employed for online performance

Table 2 Experimental design parameters and their levels

Design factor	Levels		
	1	2	3
Lubricant (*A*)	SAE20W40	Bio-lubricant	–
Compression ratio-CR (*B*)	12:1	15:1	17.5:1
Load (%) (*C*)	0	50	100

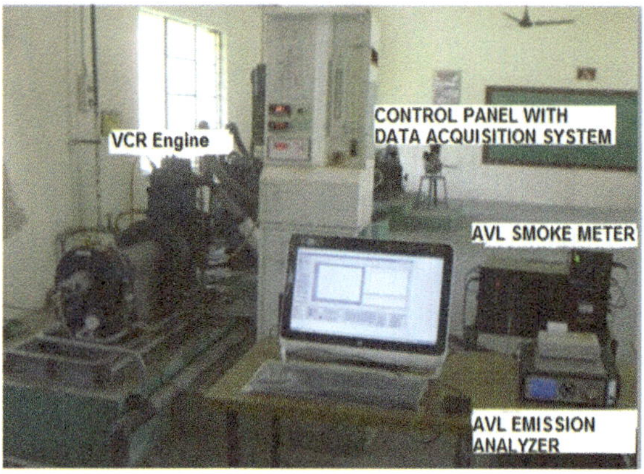

Fig. 2 Experimental setup

analysis. AVL437 smoke meter and AVL444 DI gas analyzer were used for smoke and emission measurements, respectively. The photographic view of VCR engine experimental setup is shown in Fig. 2. The specifications of test engine are summarized in Table 3. A Taguchi L_{18} orthogonal array was considered to have 18 rows corresponding to the total number of experiments.

An experiment was conducted as per the experimental design to determine performance and emission characteristics with B20 rapeseed oil biodiesel (20 % by

Table 3 Specification of VCR engine

Description	Specification
Make and model	Kirloskar/PS234
Bore × stroke	87.5 × 110 mm
Rated power	3.5 kW@1500 rpm
Compression ratio	12:1–18:1
Load indicator	Digital, range 0–50 kg, strain gauge type load cell
Dynamometer	Eddy current dynamometer
Engine type	Four-stroke, single-cylinder, direct injection, variable compression ratio diesel engine

Table 4 Experimental results

Expt. no.	Lubricant	CR	Load	Performance characteristics			Emission characteristics			
				BP (kW)	BTE (%)	ME (%)	CO (% vol.)	Smoke (HSU in %)	NO_x (ppm)	HC (ppm)
1	1	1	1	0.04	0.32	5.34	0.55	9	50	190
2	1	1	2	1.78	14.59	39.5	0.22	15	82	142
3	1	1	3	3.36	22.67	56.15	0.17	22	125	100
4	1	2	1	0.03	0.43	0.95	0.26	9	10	168
5	1	2	2	1.76	20.25	39.33	0.11	16	42	68
6	1	2	3	3.42	26.74	55.52	0.07	18	212	122
7	1	3	1	0.01	0.04	0.07	0.06	7	18	21
8	1	3	2	1.75	21.49	39.17	0.03	11	158	36
9	1	3	3	3.44	28.37	57.75	0.03	13	478	70
10	2	1	1	0.03	0.27	1.36	0.54	10	49	188
11	2	1	2	1.77	16.05	29.68	0.21	16	80	140
12	2	1	3	3.4	25.72	57.12	0.16	22	130	101
13	2	2	1	0.01	0.13	0.42	0.25	10	6	164
14	2	2	2	1.73	19.88	41.45	0.1	16	38	62
15	2	2	3	3.45	26.96	59.81	0.07	18	210	120
16	2	3	1	0.01	0.27	0.5	0.05	8	12	19
17	2	3	2	1.75	21.47	47.43	0.03	11	154	32
18	2	3	3	3.47	29.78	67.27	0.03	13	480	68

vol. rapeseed oil methyl ester +80 % by vol. Diesel) as fuel and chemically modified rapeseed oil bio-lubricant/SAE20W40 as lubricant. The experiments were repeated twice to circumvent the possible experimental errors and an average values are tabulated. In every test, a volumetric fuel consumption and exhaust gas emissions were measured and the results are summarized in Table 4.

3 Grey Relational Analysis

Grey relational analysis was employed to convert a multiple-response process optimization problem into a single-response problem with the objective function of overall grey relational grade. The corresponding level of parametric combination with a highest grey relational grade was considered as the optimum process parameter [20].

Signal-to-noise ratio (S/N) is a measure used in science and engineering for comparing the level of a desired signal to the level of background noise. Since the present study aimed at optimizing seven output response parameters, it might so happen that the higher S/N ratio for one characteristic (performance characteristics) may exhibit a lower S/N ratio for other characteristics (emission characteristics). Therefore, the overall evaluation of the S/N ratio was required for the optimization of multiple performance characteristics. The criteria for optimization of the emission characteristics were based on the smaller-the-better S/N ratio and it is expressed as

$$\mathrm{S/N} = -10\log\left[\frac{1}{r}\sum_{i=1}^{r} y_i^2\right] \tag{1}$$

The S/N ratio with a higher-the-better for the engine performance characteristic can be expressed as

$$\mathrm{S/N} = -10\log\left[\frac{1}{r}\sum_{i=1}^{r} {}^{1}\!/_{y_i^2}\right] \tag{2}$$

where y_i represents the measured value of the response variable i; r is the total number of tests. The calculated S/N ratio using Eqs. (1) and (2) is illustrated in Table 5.

Therefore, when the target value of the original sequence was "the larger-the-better," the original sequence was normalized as follows:

Table 5 S/N ratio of experimental results

Expt. no.	S/N ratio of responses in dB						
	BP	BTE	ME	CO	Smoke	HC	NO_x
1	−27.95	−9.89	14.55	5.19	−19.08	−45.57	−33.97
2	5	23.28	31.93	13.15	−23.52	−43.04	−38.27
3	10.57	27.1	34.98	15.39	−26.84	−40	−41.93
4	−30.45	−7.33	−0.445	11.7	−19.08	−44.5	−20
5	4.9	26.12	31.89	19.17	−24.08	−36.65	−32.46
6	10.68	28.54	34.88	23.09	−25.01	−41.72	−46.52
7	−40	−27.95	−23.09	24.43	−16.9	−26.44	−25.1
8	4.86	26.64	31.85	30.45	−20.82	−31.12	−43.97
9	10.73	29.05	35.23	30.45	−22.27	−36.9	−53.88
10	−30.45	−11.37	2.67	5.35	−20	−45.48	−33.8
11	4.95	24.1	29.44	13.55	−24.08	−42.92	−38.06
12	10.62	28.2	35.13	15.91	−26.84	−40.08	−42.27
13	−40	−17.72	−7.53	12.04	−20	−44.29	−15.56
14	4.76	25.96	32.35	20	−24.08	−35.84	−31.59
15	10.75	28.61	35.53	23.09	−25.1	−41.58	−46.44
16	−40	−11.37	−6.02	26.02	−18.06	−25.57	−31.59
17	4.86	26.63	33.52	30.45	−20.82	−30.1	−43.75
18	10.8	29.47	36.55	30.45	−22.27	−36.65	−53.62

$$x_i(k) = \frac{y_i(k) - \min y_i(k)}{\max y_i(k) - \min y_i(k)} \quad (3)$$

When the purpose was "the smaller-the-better," the original sequence was normalized as follows:

$$x_i(k) = \frac{\max y_i(k) - y_i(k)}{\max y_i(k) - \min y_i(k)} \quad (4)$$

where $y_i(k)$ = original reference sequence, $x_i(k)$ = sequence for comparison. i = 1, 2, …, m; k = 1, 2, 3, …, n; m, n being total no of experiments and responses. min $y_i(k)$ = smallest value of $y_i(k)$, max $y_i(k)$ = highest value of $y_i(k)$. Here, $x_i(k)$ was the value after the grey relational generation. An ideal sequence was $x_0(k)$. The grey relational grade revealed the relational degree between the experimental run sequences ($x_0(k)$ and x_i (k), i = 1, 2, …, m). The normalized S/N ratio using Eqs. (3) and (4) is illustrated in Table 6. The grey relational coefficient was calculated as

$$\zeta_i(k) = \frac{\Delta_{\min} + \Psi\Delta_{\max}}{\Delta_{oi}(k) + \Psi\Delta_{\max}} \quad (5)$$

Table 6 Normalized S/N ratio of experimental results

Expt. no.	BP	BTE	ME	CO	Smoke	HC	NO_x
1	0.238	0.314	0.631	1	0.2193	1	0.48
2	0.885	0.892	0.9225	0.6848	0.666	0.8735	0.5926
3	0.995	0.958	0.9736	0.5961	1	0.7215	0.6881
4	0.187	0.359	0.3796	0.742	0.2193	0.9465	0.1158
5	0.8838	0.9416	0.9218	0.4465	0.7223	0.554	0.441
6	0.9976	0.9838	0.971	0.29	0.815	0.8075	0.808
7	0	0	0	0.2383	0	0.0435	0.249
8	0.883	0.9507	0.921	0	0.394	0.2775	0.7413
9	0.9986	0.9926	0.9778	0	0.54	0.5665	1
10	0.187	0.288	0.432	0.9936	0.3118	0.9955	0.476
11	0.8848	0.9064	0.88	0.669	0.7223	0.8675	0.587
12	0.996	0.977	0.9761	0.5756	1	0.7255	0.697
13	0	0.178	0.26	0.7288	0.3118	0.936	0
14	0.8811	0.9388	0.9295	0.4136	0.7223	0.5135	0.418
15	0.999	0.9805	0.982	0.2913	0.824	0.8	0.8058
16	0	0.288	0.2862	0.178	0.1167	0	0.4183
17	0.883	0.9505	0.949	0	0.394	0.2265	0.7356
18	1	1	1	0	0.54	0.554	0.9932

Table 7 Grey relational coefficient of experimental results

Expt. no.	BP	BTE	ME	CO	Smoke	HC	NO_x
1	0.3961	0.7549	0.8601	0.6459	0.4506	0.4080	0.4560
2	0.8130	1.213	1.273	0.8071	0.4814	0.4135	0.4573
3	0.99	1.634	2.28	2.062	1.026	0.5423	0.4890
4	0.3808	0.7445	0.8538	0.6439	0.4503	0.4080	0.456
5	0.8114	1.21	1.269	0.8053	0.4810	0.4134	0.4573
6	0.995223	1.650	2.354	2.326	1.346	0.6636	0.5232
7	0.3333	0.7142	0.8360	0.6385	0.4493	0.4078	0.4560
8	0.8103	1.208	1.267	0.8041	0.4807	0.4134	0.4572
9	0.9972	1.657	2.384	2.4553	1.588	0.7987	0.5674
10	0.3808	0.7445	0.8539	0.6439	0.4503	0.4080	0.4560
11	0.8127	1.212	1.272	0.8068	0.4813	0.4135	0.4573
12	0.9920	1.640	2.308	2.154	1.118	0.5723	0.4970
13	0.3333	0.7142	0.8360	0.6385	0.4493	0.4078	0.4560
14	0.8078	1.204	1.261	0.8014	0.4802	0.4133	0.4572
15	0.9980	1.66	2.396	2.512	1.725	0.9032	0.6070
16	0.3333	0.7142	0.8360	0.6385	0.4493	0.4078	0.4560
17	0.8103	1.208	1.267	0.8041	0.4807	0.4134	0.4572
18	1	1.67	2.428	2.672	2.277	1.898	1.815

where $\Delta_{oi}(k) = \|X_0(k) - X_i(k)\|$ was the difference of the absolute value between $X_0(k)$ and $X_i(k)$. Δ_{min}, Δ_{max} were the minimum and maximum values of the absolute differences (Δ_{oi}) of all comparing sequences. The purpose of distinguishing coefficient ψ ($0 \le \psi \le 1$) was to weaken the effect of Δ_{max} when it became too large. In the present analysis, the value of ψ was taken as 0.5. After averaging the grey relational coefficient, the grey relational grade γ_o was being calculated.

The higher value of the grey relational grade was considered to be the stronger relational degree between the ideal sequence $x_0(k)$ and the given sequence $x_i(k)$. The ideal sequence $x_0(k)$ was supposed to be the best process response in the experimental layout. Thus the higher relational grade implied that the corresponding parameter combination was closer to the optimal. The calculated grey relational coefficient using Eq. (5) is illustrated in Table 7.

In case of engine performance, the maximum amount of brake power, brake thermal efficiency, and mechanical efficiency is observed at 100 % load (full load), Bio-lubricant, and compression ratio of 17.5. In case of emission study, the minimum amount of HC is emitted at no load, bio-lubricant, and compression ratio of 17.5; the minimum amount of CO is emitted at full load, bio-lubricant, and compression ratio of 17.5; the minimum amount of NO_x is emitted at no load, bio-lubricant, and compression ratio of 15; and the minimum smoke emission is at no load, synthetic lubricant (SAE20W40), and compression ratio of 17.5 under this combination of study. Grey relational analysis is to be performed because of multi-response data. Some of them are higher-the-better, and some of them are lower-the-better. Taguchi method cannot optimize them. The performance characteristics like brake thermal efficiency, brake power, and mechanical efficiency should be as high as possible, so the higher-the-better (HB) formula (Eq. 3) is used to normalize it; HC, CO, NO_x, and smoke emission should preferably be as low as possible, and hence the lower-the-better (LB) formula (Eq. 4) is used to normalize it.

3.1 Grey Relational Grade

All the grey relation grades are needed to be converted into "ONE" grey relation grade to perform the optimization process. The optimum combinations of engine parameters also depend on the output responses. The engine performance and exhaust emission parameters owe different weighting factors toward them as they have different impacts upon the study. The weighting factors are to be decided as per requirement. Our main goal is to substitute rapeseed oil-based bio-lubricant, yet maintaining, more or less, the same performance as synthetic lubricant. Since reduction of engine emission could be achieved by other means of emission control techniques, in general, while converting multiple grey relation grades, the value of weighting factor in engine performance was taken higher than that of emission characteristics.

However, the present investigation is aimed at running the engine with rapeseed oil biodiesel/bio-lubricant combination without any engine modification; hence, an equal weighting factor ($\beta = 0.5$) was used for both performance as well as emission characteristics [18]. When appropriate, weighting factor (β) was used with the sequence values, the general form of grey relational grade became

$$\gamma_o = \sum_{k=1}^{n} \zeta_i(k)\beta\gamma_i \ldots \tag{6}$$

Then the grey relation grade with weighting factor should be put in the response column and analyze it by the S/N ratio curve, and the S/N ratio is selected as larger-the-better as the higher value of the grey relational grade is considered as the stronger the relational degree between the ideal sequence and the given sequence. After giving this weighting factor to the individual grey relation grade, the mean grey relation grade and its rank was calculated and is given in Table 8.

Table 8 Grey relation grade and its rank

Expt. no.	Grey relational grade	Rank
1	0.6105	13
2	0.7233	7
3	0.8105	2
4	0.5114	15
5	0.6699	9
6	0.7852	4
7	0.3532	18
8	0.6339	12
9	0.7614	6
10	0.5927	14
11	0.72061	8
12	**0.8159**	**1**
13	0.4862	16
14	0.6639	10
15	0.7876	3
16	0.3847	17
17	0.6371	11
18	0.7670	5

4 Results and Discussion

4.1 Analysis of Grey Relational Grade

The signal-to-noise ratio for overall grey relation grade is calculated using the higher-the-better (HB) criteria. The results expressed in terms of S/N ratio and the mean of grey relational grades. The response to the mean of the grey relational grade is given in Table 9. The corresponding main effect plots are shown in Fig. 3. In the main effect plot, if the line for a particular parameter is nearly horizontal, the parameter has a less significant effect. On the other hand, a parameter for which the line has the highest inclination will have the most significant effect.

From the main effect plot, the parameter *C* (load) has the most significant effect among these three parameters followed by compression ratio (*B*). The optimum process parameter combination of minimum emission and better engine performance is the one which has the maximum value of signal-to-noise ratio and grey relational grade. Thus, the optimum process parameter combination is found to be $A_2B_1C_3$, i.e., bio-lubricant (*A*), compression ratio (*B*) at 12, engine load (*C*) at 100 %, i.e., full load condition.

Table 9 Mean grey relational grade-Larger-the-better

	Level 1	Level 2	Level 3
Lubricant	0.650	**0.652**	–
CR	**0.712**	0.65	0.589
Load	0.489	0.674	**0.787**

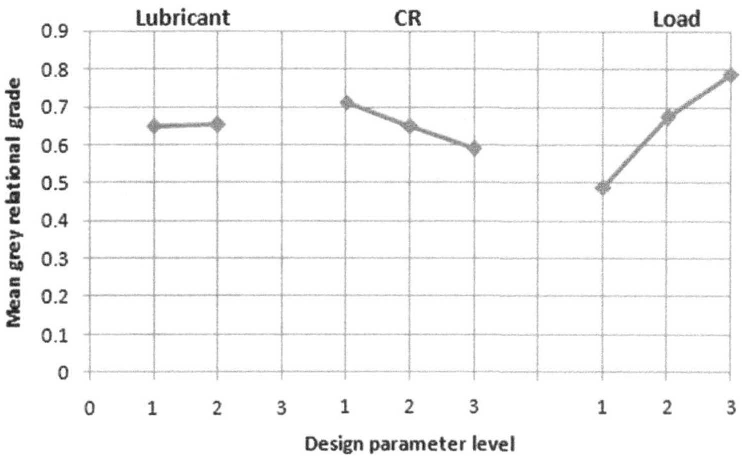

Fig. 3 Main effect plot for mean grey relational grade

4.1.1 Effect of Engine Load on Grey Grade

The effect of load on grey grade is presented in Fig. 4a. It is observed that BP, BTE, and ME are found to increase significantly with increase in load as lesser losses are encountered at higher load. The same trend is reported by Labeckas and Slavinskas with the use of biofuel/synthetic lubricant combination [21]. The main reason for this is that relatively less portion of power is lost at higher load. But it is observed that with increase in load, the smoke and NO_x emissions increase, and however CO and HC emissions decrease. With increase in load, smoke and NO_x emissions increase because a richer air–fuel ratio mixture is burnt in the cylinder which leads to an increase in combustion chamber temperature. However, at higher loads, higher temperature in the combustion chamber results in better combustion of fuel leading to very low CO and HC emissions. This result agrees with the findings of Mohanraj and Murugu Mohankumar [12]. Thus the overall engine performance is better and reflects in higher value of weighted grey relational grade at higher load.

4.1.2 Effect of Compression Ratio on Grey Grade

The effect of compression ratio (CR) on grey grade is presented in Fig. 4b. It is observed that with increase in CR, the brake power, brake thermal efficiency, and mechanical efficiency increase. This trend is observed because, at higher CR, of higher the combustion chamber temperature which results in better combustion of fuel. Moreover, the conversion of chemical energy of biodiesel fuel into mechanical energy is higher. Similarly, at higher compression ratios, HC, CO, and smoke emissions are low, because of increased combustion temperature and pressure at higher CR, and better combustion can be ensured. Also at higher CR, higher

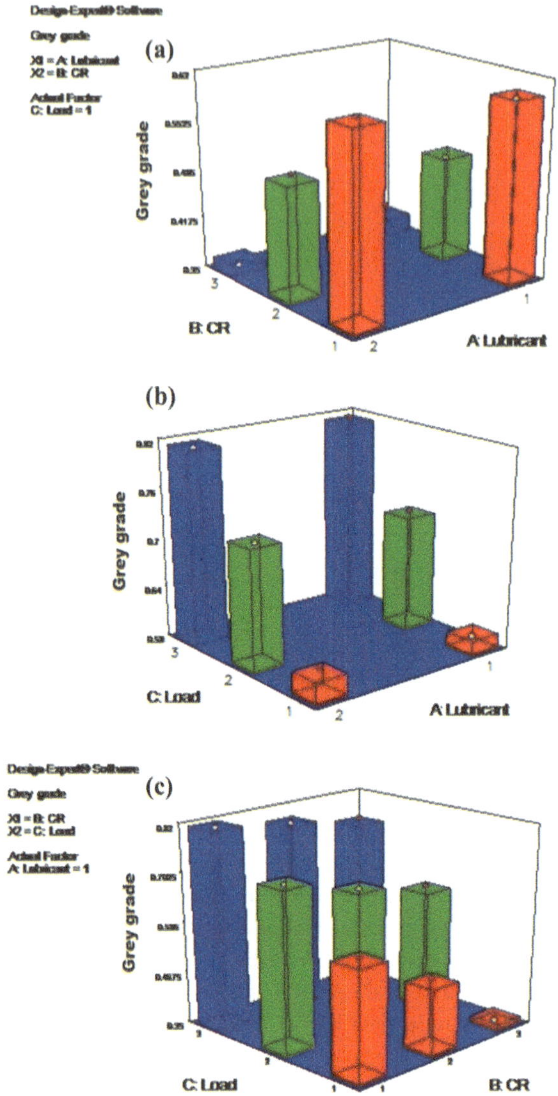

Fig. 4 Variation of grey grade on **a** CR with lubricant **b** engine load with lubricant **c** engine load with CR

combustion chamber temperature due to better combustion leads to an increase in NO_x emissions compared with lower CR. However, NO_x emissions are lower with lower CR. Therefore, the overall engine performance is better and reflects in higher value of weighted grey relational grade at lower CR.

4.1.3 Effect of Lubricant on Grey Grade

The effect of lubricant on grey grade is presented in Fig. 4c. There is not much of a variation with two different lubricants used in this study. However, the experimental results indicate a marginal improvement in BP, BTE, and ME with the use of bio-lubricant as compared with SAE20W40. This improvement in engine performance can be attributed to higher lubricity of biodiesel/bio-lubricant combination in comparison with biodiesel/SAE20W40 which reduces the frictional losses that ultimately leads to an improvement in brake power and consequently increases the brake thermal efficiency [19]. Thus, the overall performance of the engine is marginally improved when bio-lubricant is used which can be seen in the grey relational grade plot.

5 Analysis of Variance for Grey Grade

The grey relational grade obtained using Eq. 6 is analyzed using analysis of variance (ANOVA) and the analysis of means (ANOM). ANOVA is used to reveal the level of significance of influence of factors on a particular response. This is accomplished by separating the total variability of the grey relational grades, which is measured by the sum of squared deviations from the total mean of the grey relational grade into contributions by each engine parameters and the error. The response table of Taguchi method was employed here to calculate the average grey relational grade for each factor level. In this, the grouping of grey relational grades was done initially by the factor level for each column in the orthogonal array and then by averaging them. Results of ANOVA (Table 10) indicate that engine load is the highly influential parameter (percentage contribution is 80.5 %) followed by compression ratio (13.37 %). This is evident from Fig. 5a–c. It is confirmed that the improvement of mean grey relational grade is observed at full load condition as shown in Fig. 5c in comparison with no load or part load condition as shown in Fig. 4a, b, respectively.

Table 10 ANOVA results for grey grade

Factors	Degree of freedom	Sum of squares	Mean squares	F value	Prob-F	Contribution (%)
Model	5	0.3170	0.0634	37.47	<0.0001	Significant
Lubricant	1	7.28E-07	7.28E-07	0.0004	0.9838	2.15E-04
Compression ratio	2	0.0451	0.0225	13.34	0.0009	13.37
Load (%)	1	0.2718	0.1359	80.33	<0.0001	80.58
Error	12	0.0203	0.0016			6.01
Total	17	0.3373				100

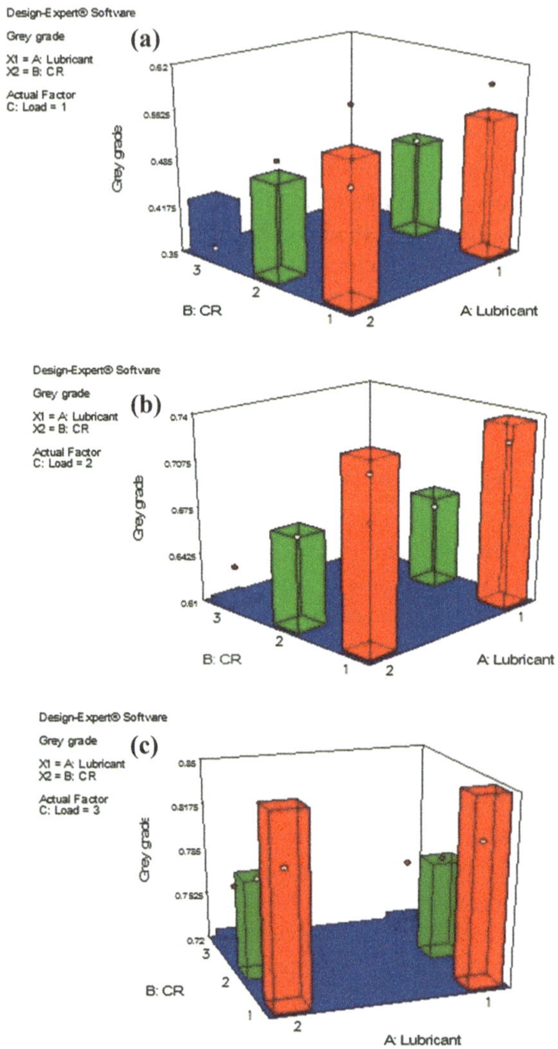

Fig. 5 Improvement of mean grey relational grade at **a** no load **b** part load **c** full load

6 Confirmatory Test

Since the optimal level of engine parameters is selected, the confirmation experiment is processed to verify the improvement of total performance characteristics. The results of the confirmation experiments are expressed by the estimated grey relational grade $\acute{\eta}_o$. The estimated grey grade $\acute{\eta}_o$ for the optimal levels of engine parameters can be calculated using Eq. (7)

Table 11 Comparison between engine characteristics using the initial and optimal level

Setting level	Initial engine parameter	Optimal engine parameter	
		Prediction	Experimental
	$A_1B_1C_1$	$A_2B_1C_3$	
BP (kW)	0.04	3.4	3.45
BTE (%)	0.32	25.72	27
ME (%)	5.34	57.12	59.8
CO (% by vol.)	0.55	0.16	0.15
Smoke (HSU)	9	22	20
NO_x (ppm)	190	130	123
HC (% by vol.)	50	101	94
Grey relational grade improvement	0.6105	0.85	0.88

$$\eta_{\text{predicted}} = \eta_m + \sum_{i=1}^{N} (\eta_0 - \eta_m) \tag{7}$$

where $\acute{\eta}_o$ is the mean of grey relational grade at optimal level. The initial design engine parameters are A_1 $B1$ C_1, which is experiment No. 1 in Table 4.

The results of confirmation experiment as given in Table 11 indicate that the brake power is increased from 0.04 to 3.4 kW, the brake thermal efficiency is improved from 0.3 to 25.7 %, the mechanical efficiency is improved from 5.3 to 57 %, HC emission is decreased from 190 to 101 ppm, CO emission is decreased from 0.55 to 0.16 % by vol., smoke emission is increased from 9 HSU to 22 HSU, and NO_x emission is increased from 50 to 130 ppm. The estimated grey grade increases from 0.6105 to 0.85, which is the largest value obtained in all experimental results as given in Table 8. It is found that there is an improvement of 5 % observed in the weighted grey relational grade. This ensures the usefulness of grey relational approach to the process optimization, where multiple quality criteria have to be fulfilled simultaneously.

7 Conclusions

The outcome of the present investigation is as follows:

- The VCR engine can safely be operated at various compression ratios with the formulated rapeseed oil bio-lubricant without any engine modifications and significant changes in engine power and fuel economy.
- Even the marginal improvement in BP, BTE, and ME with the use of rapeseed oil bio-lubricant as compared to SAE20W40. This can be attributed to the higher lubricity of rapeseed oil-based bio-lubricant.

- Among the tested parameters, the engine load has the strongest correlation to the engine performance and emission characteristics. The recommended levels of engine parameters for maximizing the engine performance and minimizing the engine exhaust emissions simultaneously are lubricant at level 2 (bio-lubricant), compression ratio at level 1 (CR-12), and engine load at level 3 (100 % load-full load), which shows that the use of rapeseed oil-based bio-lubricant improves the performance in VCR engine fueled with rapeseed oil biodiesel.
- An increase in the value of predicted weighted grey relational grade from 0.6105 to 0.88 confirms the improvement in the engine performance of VCR engine using optimal values of engine parameters.

From this detailed study, it can be concluded that vegetable oil-based bio-lubricant is a strong renewable candidate for the replacement of petroleum-based lubricant in the near future, especially the developing country like India, as India is an agricultural-based country.

References

1. Demirbas A (2009) Biodegradability of biodiesel and petrodiesel fuels. Energy Sour—Part A 31:169–174
2. Arumugam S, Sriram G (2012) Effect of bio-lubricant and bio-diesel contaminated lubricant on tribological behavior of cylinder liner-piston ring combination. Tribol Trans 55:438–445
3. Ssempebwa JC, Carpenter DO (2009) The generation, use and disposal of waste crankcase oil in developing countries: a case for Kampala district of Uganda. J Hazard Mater 161:835–841
4. Mercurio P, Burns KA, Negri A (2004) Testing the eco toxicology of vegetable versus mineral based lubricating oils: degradation rates using tropical marine microbes. Environ Pollut 129:165–173
5. Uosukainen E, Linko YY, Lamasa M, Tervakangas T, Linko P (1998) Transesterification of trimethylol propane and rapeseed oil methyl ester to environmentally acceptable lubricants. J Am Oil Chem Soc 75:1557–1563
6. Asadauskas S, Erhan SZ (1999) Depression of pour points of vegetable oils by blending with diluents used for biodegradable lubricants. J Am Oil Chem Soc 76:313–316
7. Kim JR, Sharma S (2012) The development and comparison of bio-thermoset plastics from epoxidized plant oils. Ind Crops Prod 36:485–499
8. Ting C-C, Chen C-C (2011) Viscosity and working efficiency analysis of soybean oil based bio-lubricants. Measurements 44:1337–1341
9. Arumugam S, Sriram G, Rajmohan T (2014) Multi response optimization of epoxidation process parameters of rapeseed oil using response surface methodology (RSM) based desirability analysis. Arab J Sci Eng 39:2277–2287
10. Arumugam S, Sriram G (2013) Synthesis and characterization of rapeseed oil bio-lubricant—its effect on wear and frictional behaviour of piston ring-cylinder liner combination. Proc IMechE Part J: J Eng Tribol 227:3–15
11. Arumugam S, Sriram G (2014) Synthesis and characterization of rapeseed oil bio-lubricant dispersed with nano copper oxide—its effect on wear and frictional behaviour of piston ring-cylinder liner combination. Proc IMechE Part J: J Eng Tribol. doi:10.1177/1350650114535384
12. Mohanraj T, Murugu Mohankumar K (2013) Operating characteristics of a variable compression ratio engine using esterified tamanu oil. Int J Green Energy 10:285–301

13. Agarwal AK (2005) Experimental investigation of the effect of biodiesel utilization on lubricating oil tribology in diesel engines. Proc IMechE Part J: J Eng Tribol 219:703–713
14. Mumtaz MW, Adnan A, Mahmood Z, Mukhtar H, Malik MF, Qureshi Fahim Ashraf, Raza Ahtisham (2012) Biodiesel from waste cooking oil: optimization of production and monitoring of exhaust emission levels from its combustion in a diesel engine. Int J Green Energy 9:685–701
15. Karnwal A, Hasan MM, Kumar N, Siddiquee AN, Khan ZA (2011) Multi-response optimization of diesel engine performance parameters using thumba biodiesel-diesel blends by applying the Taguchi method and grey relational analysis. Int J Automot Technol 12:599–610
16. Maheshwari N, Balaji C, Ramesh A (2011) A nonlinear regression based multi-objective optimization of parameters based on experimental data from an IC engine fueled with biodiesel blends. Biomass Bioenergy 35:2171–2183
17. Alonso JM, Alvarruiz F, Deantesjm Hernandez L, Hernandez LV, Molto G (2007) Combining neural networks and genetic algorithms to predict and reduce diesel engine emission. IEEE Trans 11:46–55
18. Wu H-W, Wu Z-Y (2013) Using Taguchi method on combustion performance of a diesel engine with diesel/biodiesel blend and port-inducting H_2. Appl Energy 104:362–370
19. Arumugam S, Sriram G, Ellappan R (2014) Biolubricant—biodiesel combination of rapeseed oil: an experimental investigation on engine oil tribology, performance and emission of variable compression engine. Energy 72:618–627
20. Kuo Y, Yang T, Huang GW (2008) The use of grey based Taguchi method to optimize multi response simulation problems. Eng Optim 40:517–528
21. Labeckas G, Slavinskas S (2005) The effect of diesel fuel blending with rapeseed oil and rapeseed oil methyl ester on engine performance and exhaust emissions. J KONES Int Combust Eng 12:1–12

Chapter 5
Biolubricants and the Potential of Waste Cooking Oil

J.G. Alotaibi and B.F. Yousif

Abstract In the current decade, development of recycle, renewable, and sustainable products to replace fossil products is an essential and important matter from industrial, environment, and academic point of views. The excessive usage of Petroleum-based oils significantly contributes to the pollution of the environment and had caused environmental pollution and awareness from the environmental sectors. Researchers start exploring an alternative oil from natural resource aiming to replace the fossil oil and this becomes the main ambitious of many researchers, environmental, and government bodies. In this chapter, a comprehensive literature review is introduced and several issues are addressed with regards of the usage of newly developed lubricants that are based on vegetable oils. Furthermore, it is exploring the potential of using waste cooking oil as lubricant for tribological applications.

1 Introduction

Later in the 1800s, petroleum had been discovered and that led to the replacement of animal fats, vegetable oils, and mineral oils with synthetic oils. Petroleum oil had gradually started to be the main lubricant base stocks, and that was because of their low cost and superior performance. Lubricants are being used widely in all fields of manufacturing and industrial applications. Studies showed that more than thirty eight million metric tons of oils were used in lubrication techniques in 2005 for different industrial applications in the United States (USA). Lubricants are

J.G. Alotaibi
Department of Automotive and Marine Technology, The Public Authority for Applied Education and Training, College of Technological Studies, Kuwait, Kuwait

B.F. Yousif (✉)
School of Mechanical and Electrical Engineering, Faculty of Health, Engineering and Sciences, The University Southern Queensland, Toowoomba, QLD 4350, Australia
e-mail: Belal.Yousif@usq.edu.au

J.P. Davim (ed.), *Ecotribology*, Materials Forming, Machining and Tribology,
DOI 10.1007/978-3-319-24007-7_5

commonly used to reduce overheating and friction in a variety of engines, machinery, turbines, and gear. The excessive usage of petroleum-based oils significantly contributes to the pollution of the environment [1] and had caused environmental pollution and awareness from the environmental sectors. Besides that, the demand for fossil fuel and oil products is increasing in numerous areas. From the reported works, alternative oil should increase to cover about 36 billion gallons in 2022 [2]. The literature showed that bio-oil becomes the most successful candidate for biofuel since, in the current decade, there are few attempts aiming to study the potential of using bio-oil such as sunflower oil [3, 4], castor oil [5, 6], soybean oil [7, 8], etc. as biofuels for diesel engines. Most of the works showed good and promising results. However, there is a tribological issue raised by most of the researchers in which biofuels deteriorate the engine components.

Besides the usage of the bio-oil as fuel, there is an effort that is currently put to use bio-oil as lubricant for several tribological applications [9–11], e.g., soybean oil (USA and South America) [12], rapeseed oil (Europe) [13], and palm oil (Asia) [14–16]. The studies are still in the initial stage and there are many issues and limitations need to be addressed [17, 18]. Moreover, the literature highly recommends deep investigation to study the performance and the potential of using biolubricants [19–21].

Lubricants are being divided into solid, semisolid, gas, or fluid. Lubricants are classified into two major groups: (i) automotive lubricants and (ii) industrial lubricants. Automotive lubricants have to perform in different types of vehicles both petrol and diesel under a variety of operating conditions. Quality requirements of such lubricants are established by the Society of Automotive Engineering (SAE) and are specified in its classification system [22]. Industrial oils are used for protecting internal elements from intensive cycles of operation. Viscosity of industrial oils is an essential factor to maintain operating at an optimal condition over a wide range of temperature and working conditions. Industrial lubricants are classified by the International Standards Organization (ISO) from the American Society for Testing and Materials (ASTM) D2422. The ISO classification is based solely on viscosity ranges at 40 °C. In the light of the above, the current paper is motivated to cover the recent literature on the biolubricants and explore the potential of using waste cooking oil as lubricant for tribological applications.

2 Lubricants and Biolubricants

Tribology science covers friction, wear, and lubricant branches. The science of tribology starts long time ago about 4500 BC when the first wheel has been developed [23]. This followed by animal fat to ease the rock movements in building Egyptian pyramids. A lubricant is used to be a protector between devices. When two devices are in contact with each other, a contacting pressure will be created between them causing surface damage if there was not any protector between them. Lubricant is used to reduce and lower the wear between the components of any mechanical devices [24]. 100 years ago, water was mainly used to cool cutting tools due to its high availability and thermal capacity. Lubrication and cooling in machines are vital to reduce the effect of any cutting process at the interface of a cutting tool–workpiece. The main drawbacks in the coolant are the poor lubrication and corrosion of the machines. On the other hand, mineral oils were used at that time due to their higher lubricity, but the high costs and low cooling ability of them had led to only using them in "low cutting speed machining operations" [25]. Any lubricant can also be used to remove and reduce the heat from a device during its operation. Lubricants are commonly used to reduce overheating and friction in a variety of engines, machinery, turbines, and gear [26, 27]. The development of tribology science continues since that time until now. With the discovery of fossil oil, new areas of research have been developed owing to discover the synthetic lubricants for various tribological applications. Lubricants are being used widely in all fields of manufacturing for lubricating their materials and machines. From economical point of view, more than thirty eight million metric tons of oils were used for lubrication techniques. In the current era, petroleum-based lubricants are mostly used in industries. However, the excessive usage of such lubricants is effecting the environment leading to numerous issues related to environments such as groundwater and surface water contamination, soil contamination, and air pollution [1, 28, 29]. Specifically, synthetic lubricants can be emitted into the environment through cleaning activities and accidental leakage. Further to this, wastewater lubricants include free oil and emulsified oil, created by a mixture of oil with a wastewater and washing agent. Several techniques have been developed to overcome such problems such as corrugated plate interception [30] or gravitational oil separation [31].

Recently, the excessive world consumption of crude oil has several impacts to the world in term of economic since there is a huge increase in the price of the fuel products. Moreover, there is a huge impact on the environment since there are challenges from global and environmental sectors. This encourages the researchers in the area of oil and fuel products to find alternative friendly product to replace the synthetic product for several applications [32]. In addition to that, this alerts the reconsideration of using friendly lubricants with renewable properties and sources, like nonedible vegetable oils [33]. The high interest in the research area of the biolubricants is evidence in the increase of number of publications related to the area in sciencedirect database Fig. 1. Due to the increase interest in the research area of biolubricants, the following sections will focus and discuss the recent issues raised by the reported works to cover the potential biolubricants and the required characteristics of the lubricants.

2.1 Biolubricants in the Recent Era

Most of the lubricants used in rotary machine elements are based on mineral and synthetic oils combined with different additives to meet the requirements of the

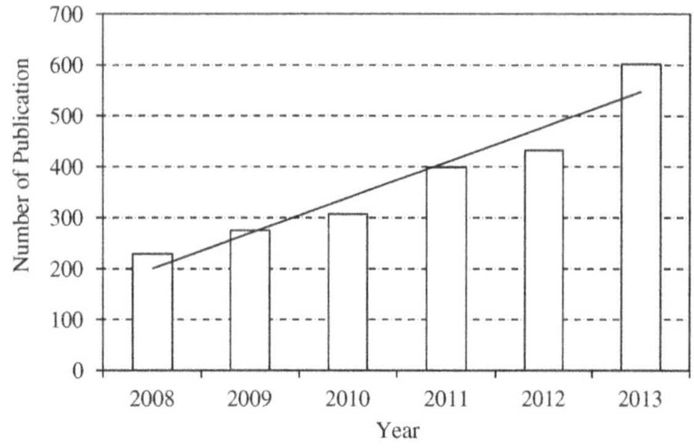

Fig. 1 Extracted data from international database (www.sciencedirect.com) showing the number of publications in the area of biolubricants

intended applications [34]. The main disadvantages and limitations of mineral oils are the poor biodegradability, high cost, and limited resources [35], i.e., there is a need for substitutions of friendly environmental lubricants [36] due to strict government and environmental regulations as reported by recent work [19, 37]. Petroleum-based lubricants are toxic to environment and difficult to dispose, and many lubricant manufacturers have reconsidered vegetable oils over synthetic fluids due to a combination of renewability, biodegradability, excellent lubrication performance, and low cost [38, 39].

Bio-oil can be divided based on its resources into vegetable and animal. Vegetable oil is more popular and common compared to the animal due to its availability, ease in extraction, and low cost [40, 41]. Vegetable oils are promising candidates as base fluid for eco-friendly lubricants. From the reported works, vegetable oils as lubricants have numerous advantages such as good contact lubrication [42], excellent lubricity [40], biodegradability, viscosity–temperature characteristics [43, 44], very low volatility high viscosity indices as minimum changes in viscosity with temperature [45], and high flash point due to the high molecular weight of the triglyceride molecule and excellent temperature–viscosity properties [46].

Despite of the above advantages, vegetable oils can be edible or nonedible oil. Edible oils are more common and available compared to the nonedible oils. However, since most of the vegetable oils are edible, a limitation can be found in using such oil in tribological applications. Therefore, it is not highly recommended to use edible oil as lubricant and/or either fuel [37, 47, 48]. In this scenario, there are two demands that need to be considered in developing a new oil which should have less impact on the environment and be nonedible. The researcher of this project finds that waste cooking oil is a potential candidate as alternative to the synthetic and edible oils. From fuel point of view, there are several researches that are going on in the current year aiming to convert waste cooking oil into biofuels using different techniques as reported by [49, 50]. In the next section, a summary of the current works on waste cooking oil that is introduced.

Table 1 Different types of vegetable oils and their applications

Vegetable oils	Applications
Canola oil [51]	Hydraulic oils, metalworking fluids, "food grade lubes," penetrating oils, "chain bar lubes tractor transmission fluids"
Castor oil [33]	Greases, gear lubricants
Coconut oil [52]	"Gas engine oils"
Crambe oil [53]	Intermediate chemicals, grease, surfactants
Cuphea oil	Motor and cosmetic oils
Jojoba oil	Cosmetic industry, grease, lubricant applications
Linseed oil	Paints, coating, lacquers, stains, varnishes,
Olive oil [51]	Automotive lubricants
Palm oil [54]	Grease, steel industry, rolling lubricant,
Rapeseed oil [33]	"Air compressor-farm equipment," "chain saw bar lubricants," biodegradable greases

Vegetable oils are promising candidates as base fluid for eco-friendly lubricants because of possession of some excellent properties for their potential use as a base stock for lubricants and functional fluids such as good contact lubrication, excellent lubricity, biodegradability, viscosity–temperature characteristics, very low volatility high viscosity indices (VI) (i.e., minimum changes in viscosity with temperature), and high flash point due to the high molecular weight of the triglyceride molecule and excellent temperature–viscosity properties. On the other hand, VOs in its natural form cannot fully meet the performance criteria for the most lubricants due to their drawbacks including poor low-temperature properties such as opacity, precipitation, and poor flowability at relatively moderate temperature. Table 1 demonstrates possible applications for different vegetable oils.

With regard of the vegetable oils, most of the works have been studying the wear and frictional performance of the oil without any chemical additives as reported recently by Madankar et al. [55] on castor seed oil. Castro et al. [56] studied the wear properties of different modified neat soybean oils without any additives. In that work, the test has been performed using tribological setup at fixed speed of 700 rpm. There is no remarkable effect of the different oils on the wear and frictional performance of the soybean oil. Table 2 summaries the recent works on

Table 2 Main physical properties of vegetable oils [54]

Vegetable oils	Palmitic (16:0)	Stearic (18:0)	Oleic (18:1)	Linoleic (18:2)	Linolenic (18:3)	Linolenic (18:3)	Unsaturated/saturated ratio
Castor oil	2.63	1.51	4.74	8.36	–	82.80	23.20
Soybean oil	11.28	2.70	24.39	56.28	5.34	–	6.15
Rapeseed oil	4.56	–	65.99	21.13	8.16	–	20.90
Sunflower oil	6.18	2.16	26.13	65.52	–	–	11.00
High-oleic sunflower oil	3.84	4.42	83.66	8.08	–	–	11.10

vegetable oil as lubricant and their findings. In general, there is interest and high attention from the researchers to focus their studies on the possibility of using vegetable oil as lubricant for several applications. However, there is no clear direction on the application of vegetable oil yet since such oil is new and need deep investigation.

In [12], mixtures of the original soybean oil, the epoxidized soybean oil, and the hydrogenated soybean oil as the base oils have been examined to determine the viscosity and working efficiency of the oils. The applications of those mixtures of oils have focused for internal combustion engines. The results showed that the epoxidized soybean oil has extremely large viscosity in comparison with the engine lubricants as well as the original soybean oil, whereas the hydrogenated soybean oil is clearly opposite. This viscosity analysis offers good information to fit viscosity of the engine lubricants by mixing the three soybean oils as base oils. For the two-stroke engines, castor oil-based lubricants have been used owing to reduce the smoke and emission level, in comparison with the conventional oil used for two-stroke engines (*2T-Lubricant*, [57]).

2.2 Characteristics of Biolubricant Oils

Low resistance to oxidative degradation and poor low-temperature properties are the main performance issues accounted in using vegetable oils as lubricant oils. Therefore, there are different methods to solve these problems such as reformulation of additives, chemical modification of vegetable-based oils, and genetic modification of the seed oil crop. Triethanolamine oleate, triethanolamine, and oleic acid are the main additives of the base oils which are used as lubricant oils. These additives affect the thermal stability of rapeseed and tribological behavior of base oils. The additives demonstrated significantly better thermal stability and tribological behavior. Also, the thermal stability of vegetable oils can be enhanced by chemical modification techniques. Modifications of the "carboxyl group" and modifications of the "fatty acid chain" are the two main chemical modification techniques. Modifications of the carboxyl group include Esterification/Trans esterification techniques and modifications of the fatty acid chain include selective hydrogenation (Dimerization/oligomerisation), formation of C–C and C–O bonds, metathesis, and oxidation techniques. The different methods under genetic engineering are used and new vegetable oil types are being used. Sunflower and "high oleic soybean oils" are some good examples. These oils have higher "thermo-oxidative stability" and higher load transportation capacity. These oils need less adjustment to be used as "base oil lubricants," compared to conformist plant based [33].

"Low temperature performance" is one of the main limitations for the usage of vegetable oils as lubricants, more than synthetic oil-based or mineral lubricants. [54] studied the low-temperature behaviors of different types of vegetable oils which are used in lubricating applications. Also, [54] studied vegetable oils' behaviors after blending them with "pour point additives." Blends are prepared by rotating the samples at 300 rpm at 100–150 °C for 5–10 h, depending on concentration of additives and types of them. This thermal process is required to make sure that the additives are completely soluble in vegetable oils. After that, samples are cooled at room temperature. Different vegetable oils are used as base stocks such as castor (CO), soybean (SYO), rapeseed (RO), sunflower (SO), and high-oleic sunflower (HOSO). CO oils are received from Spain; SYO, RO, and HOSO oils are supplied by Germany and Spain; and SO oils are obtained from local supermarket. Main physical properties of vegetable oils are demonstrated in Table 2.

Thermal analysis by "Differential scanning calorimetry" (DSC), "Pour point temperature measurement," "Viscosity measurements at low temperature," and Statistical analysis are the main methods used to study the behavior of vegetable oils. Thermal analysis by "Differential scanning calorimetry" (DSC) can be defined as the analysis of cooling curves (heat flow (W/g) vs. temperature) and their blends with cold flow and viscosity improver additives which are obtained using a differential scanning calorimeter (Q-100). In this method, samples are heated in hermetic aluminum tubes at 25 °C, and directly cooled with a cooling rate of 5 °C/min to −80 °C. Also, samples are cleaned with nitrogen which has a flow rate of 50 mL/min and then to determine freezing temperature and wax appearance, the cooling curve for each sample is analyzed. The pour point can be defined as the lowest temperature for vegetable oil when it is cooled. The pour point of vegetable oils and their blends with additives are determined by (Standard Test Method for Pour Point of Petroleum Products (ASTM D97-02)). In viscosity measurements at low-temperature method, the dynamic viscosities for vegetable oils will be measured using coaxial cylinder in "rotational controlled-strain rheometer" [54].

"Differential Scanning Calorimetry" (DSC) is a method used to determine the crystallization for "vegetable-based lubricants." Also, DSC method is more accurate and faster than viscosity measurements and pour point temperature at low-temperature methods. In all tests, castor oil demonstrates a better behavior at low temperature, because it has low content of saturated fatty acids and it has hydroxyl groups in the fatty acid chain which can obstruct the crystal packing system of triacylglycerols (TAG) molecules. Vegetable oils which have lower ratio of "unsaturated/saturated fatty acids" crystallize at higher temperatures. Additionally, the concentrations of "Polyunsaturated Fatty Acids" (PUFAs) in vegetable oils have more impact on low-temperature properties than the concentration of "saturated fatty acids." Therefore, the "rapeseed oil" has better behaviors at low temperatures compared with other types of oils which have similar molecular structure such as HOSO, SO, and SYO oils. Generally, "The Pour Point Depressant" (PPD) additives are used to improve the low-temperature behaviors of vegetable oils. PPD additives increase the low-temperature performance and decrease the pour point for vegetable oils which depend on fatty acid composition in vegetable oils. The results demonstrate that the blend of sunflower and the pour point depressant (SO/PPD) has lower pour point than neat oil (HOSO/PPD) [54].

Biodegradability investigation of lubricants using standardized tests used to provide valuable information for regulation assessment and purposes of how chemical structure of lubricants influences biodegradability. Poor solubility of lubricating base oils in water is the major problem which obstructs the biodegradability analysis. Ultimate and primary biodegradability are the two phases of biodegradability which are used for analyzing different chemical structure oils such "syntheticpolyolester" oils, rapeseed oil, conventional mineral oils, and poly (a-olefin) oils. "Co-coordinating European council for the development of performance tests for lubricants and engine fuels" (CEC L-33-A-93 test) is used to evaluate primary biodegradability of lubricants. "Organization for Economic Cooperation and Development Guidelines for Testing of Chemicals" and "301B Ready biodegradability" (OECD 301 B and OECD 310 tests) are used to evaluate ultimate biodegradability of lubricants. Primary biodegradability is evaluated according to the CEC L-33-A-93 test using triplicate flasks containing "Di-IsotridecylAdipate" (DITA) as the reference material, and triplicate flasks contain the test oils, "duplicate neutral flasks" and "duplicate poisoned flasks," which are prepared for various durations of time (0, 7, 14, and 21 days) during the test [55]. Despite the low-temperature properties and oxidative stability of vegetable oil-based lubricants compared with petroleum-based lubricants, it is used in many countries. For example, in USA people use corn oil and soybean oil, while in Europe and North America rapeseed oil is used [9].

2.3 Edible and Inedible Oils

Recently, there is a big concern about using edible vegetable oil or the feedstock first generation, because it may cause starvation in poor and developing countries, the other problem appears in utilizing the available "arable land," and it can create ecological imbalances when countries start cutting forests. Hence, these feedstock cause deforestation and wildlife damage. Therefore, second-generation feedstock or "non-edible vegetable oils" will be attractive to produce biodiesel, besides for the "sustainable production of biodiesel" the second-generation feedstock is very promising. There are examples of nonedible seed crop oils like jatrophacurcas, tobacco, deccan hemp, castor, jojoba, sea mango, coriander, salmon oil, desert date, cardoon, milkweed, tung, and lucky bean tree. Microalgae oils are considered to be "inexhaustible source of biodiesel." They are very economical when compared with edible oils. Microalgae give the highest oil yield, and its yield is "25 times higher than the yield of traditional biodiesel crops." Waste of cooked vegetable oils is another biodiesel feedstock with cheap price relatively for production of biodiesel from fresh vegetable. As an economical source biodiesel production from cooked vegetation oil is good option, global consumption of biodiesel feedstock should rely on multiple sources since expiring one source will bring harmful effects on the long term, and biodiesel feedstock should be as diversified as possible, depending on geographical locations in the world [7].

Using noneditable vegetables will not contradict with countries demands for food. Vegetable oil lubricants cover a small market segment and rise slowly and steadily in open applications such as chainsaws, forestry, two-stroke engines, etc. More efforts had been taken to change the global laws and policies to ensure environmental safety, and all cases of lubricants interference environmental compatibility must be checked. Nonedible vegetable oils have the potential to divert the agricultural practices and strengthen the economics [5].

Biomass like edible, nonedible crops, microorganisms, algae, recycled cooking greases, wood (lignocelluloses), and animal waste are used to derive bio-oils. The most popular virgin crops used for the production of bio-oil are canola, corn, soybean, rapeseed, mahua, mustard, jatropha, safflower, sunflower, and palm. For bio-ethanol production, sweet sorghum, straw, sugarcane/beet and, rice, wheat, and corn can be used. The debate of fuel versus food and the environmental impacts concerned about conversion and cultivation will limit using food crops to produce fuel. Converting residues from lignocellulose material or wood into biofuels is difficult; in the next decades advanced technology is expected to "reach their commercial stage," which does not require lands to produce algae, fungi, yeast, and bacteria to reach 70 %. However, production of large scale of oil from microorganisms and microalgae can be a challenge [8].

3 Waste Cooking Oil

In using edible bio-oils as fuels, several feedstock have been proven impractical or infeasible because of their extremely high cost due to their usage primarily as food resources as reported recently by Yaakob et al. [50]. Waste cooking oil (WCO) can be considered the most promising bio-oil feedstock despite its drawbacks, such as its high free fatty acid (FFA) and water contents [50, 58]. Waste cooking oil (WCO) can be produced from different sources and its base materials are plant-based lipids (sunflower, corn, margarine, coconut, palm, olive, soybean, oil, and canola) or animal-based lipids (butter, ghee). It can be freely collected from food production industries, restaurants, and houses using a special "recycle bin," this requires public awareness, [50]. WCO and fats produce significant disposal problems in many parts of the world and the growing problem of wastes affects the daily lives of millions of people, [59, 60] who reported that the estimated amount of WCO collected in Europe is about 700,000–100,000 tons/year. In United States, there is about 11 billion liters (2.9 billion gallons) of recycled vegetable oil produced yearly which is mainly from industries of deep fryers, snack food, and fast food [61].

To make use of such waste oil, several studies have been conducted aiming to convert the oil into biofuels especially in the current era as reported by [49, 62]. Table 3 lists waste cooking oil applications in the recent era. As reported by many researchers, biofuels produced from waste cooking oils have numerous advantages such as low pollution (CO2, CO, and NOx), low cost, and acceptable brake-specific fuel consumption. Moreover, biodiesel produced from waste cooking oil may improve the pump plunger lubrication conditions [63]. However, it has been

Table 3 Waste cooking oil application in the recent era

References	Oil type	Applications	Remarks
Bio-oil from the pyrolysis of palm and Jatropha wastes in a fluidized bed [67]	Palm kernel shell		The finding indicates that they were similar to palm fatty acid distillate from palm oil and could be used as alternative feedstocks for biodiesel production using hydrotreating process
Bio-oils from microwave pyrolysis of agricultural wastes [68]	Sago wastes		The bio-oils have a potential as valuable source for fuel or chemical feedstocks
Performance, emission, and combustion characteristics of a variable compression ratio engine using methyl esters of waste cooking oil and diesel blends [69]	Waste cooking oil methyl ester	Single cylinder four-stroke engine	The blends when used as fuel results in reduction of carbon monoxide, hydrocarbon, and increase in nitrogen oxides emissions
[58]	Waste cooking oil (sunflower oil)	As fuel for diesel engine	Showed promising results with some optimization
[70]	Collected from fish restaurant	As fuel for diesel engine	In this comparative study, conversion of waste cooking oil to methyl esters was carried out using the ferric sulfate and the supercritical methanol processes. This process resulted in a feedstock to biodiesel conversion yield of about 85–96 % using a ferric sulfate catalyst

reported that biodiesel produced from waste cooking oil highly damages the engine components [64]. In general, biodiesels have significant effects on the engine from tribological point of view, i.e., some biodiesel properties such as higher viscosity, lower volatility, and the reactivity of unsaturated hydrocarbon chains which causes injector coking and trumpet formation on the injectors, more carbon deposits, oil ring sticking and thickening, and gelling of the engine lubricant oil [63]. Many biodiesel investigations are related to wear of engine components such as piston, piston ring, cylinder liner, bearing, crankshaft, cam tappet, valves, and injectors [65, 66].

Besides the above needs to make use of waste cooking oil and its influence on the engine life, an interest can be drawn to use waste cooking oil as lubricant. Based on the knowledge of recent publications, there is no research work that has been attempted in using waste cooking oil as lubricant. However, there are several articles and industrial applications which have been reported on using vegetable oils as lubricants. The edibility of vegetable oil is the main limitation in using it as lubricants. Therefore, waste cooking oil can be used as lubricant for better application compared to biofuels and/or using virgin vegetable oils as lubricants.

3.1 *Operating Parameters Effects on Wear and Frictional Behavior of Wet Contact Surfaces*

There are several works that have been reported in studying the frictional and wear behavior of metals under neat biolubricant conditions. Up to the recent time, most of the works focused on specific application in which the sliding speed, sliding distance, environmental temperature, and applied load are fixed, this can be found in the work reported recently by Sharma et al. [71] and Luo et al. [18]. However, it is well known that the operating parameters have significant influence on the wear and frictional behavior of metal contact under dry and/or lubricant conditions as reported by [72–74]. For example, in [73], friction and wear behavior of gray cast iron has been investigated under different test parameters, i.e., applied load, sliding speed, and test environment. He found the followings:

1. The wear loss increased with increasing sliding speed and load. However, the presence of the oil lubricant caused a reduction in the wear loss.
2. Temperature near the specimen surface increased with test duration. The rate of increase was high initially followed by a reduction in the rate of increase. Friction coefficient also followed an identical trend in general.
3. Increasing applied load and sliding speed brought about higher frictional heating while the severity of heating reduced in the presence of the oil lubricant.
4. The friction coefficient decreased with increasing load, wherein the rate of reduction was high initially followed by the attainment of a steady-state value. Also, increasing sliding speed caused the friction coefficient to decrease during dry sliding while it produced a mixed effect on the property in the presence of the oil lubricant.

In other words, the operating parameters should be considered in testing a new lubricant. This motivates the current study to focus on the influence of the operating parameters on the newly developed bio-oil from waste cooking oil.

3.2 *Effect of Lubricant Temperature on the Wear and Frictional Behavior of Metals*

With regard of the biolubricants, several works that have been conducted to study the influence of the oil temperature on the viscosity of the bio-oil in which the oil was heated up and the viscosity was determined. For instance, Ting and Chen [12] studied the viscosity of soybean oil-based biolubricants at different temperatures and they found that increasing the temperature drops, the viscosity of the oil which in turn affects the friction and wear characteristics of the rubbed surfaces. Although vegetable oils have some excellent properties for their potential use as lubricants, some inconveniences should be technologically improved, i.e., limited range of viscosities available. Quinchia et al. [45] used ethylene–vinyl acetate copolymer as viscosity modifier for sunflower oil (SO), high-oleic sunflower oil (HOSO), and soybean oil (SYO). The viscosity experiments have been conducted at moderated temperature (below 40 °C) and the copolymer found to be very effective since it highly improves the viscosity of the oil. Similar works have been reported by many researchers such as [22, 39, 55]. Those works focused on the influence of the temperature on the oil viscosity only where in practical application, the oil will be in the rubbing area and the interface temperature associated with the environmental temperature may have combine effect on the friction and the wear behavior. In this work, the research focuses on the combination of both the environmental and the interface temperatures since a new tribology machine will be developed to conduct such experiments, see the new tribology machine section.

4 The Potential of Using Waste Cooking Oils as Lubricants

Viscosity is one of the most important properties used to determine and choose the suitable lubricant. In the case that the viscosity of any lubricant is high, the lubricant requires a larger force to overcome its intermolecular forces. When viscosity of the lubricant is very low, then the surfaces between the devices are rubbed leading to damage to the components. In this article, the main focus is to determine the potential of using waste cooking oil from viscosity point of view. In the following sections, the experimental procedure with the preliminary results will be presented.

4.1 Oil Collections and Experimental Procedure

There are different resources for bio-oil. In the current study, waste cooking oil was collected from a restaurant in Toowoomba city which is used for fish and chips production. In the preparation and cleaning process of the lubricant, the collected waste cooking oil was first filtered using sieves to remove the big undesired materials. This is followed by heating up the oil to a temperature of about 80 °C and then filtered using microfilter, which was supplied by Sefara PLY filter. Different blends of lubricants were prepared by mixing the prepared waste cooking oil with fully industrial synthetic oil (10 W-40). Viscosity of the oil was tested using ISL Viscometer at. In the current study, the viscosity of the oil is compared with the fully syntactic oil. Kinematic and dynamic viscosities of the prepared oils were measured according to the ASTM D 445 (ASTM Standards, 1991b) and were carried out at 40–100 °C.

4.2 Initial Results

Dynamic viscosity of the collected oil and its blends against different temperatures is plotted in Fig. 2a, b in two forms. It has been found that some of the data in the literature are presented in log form, [12] and for comparison purposes, the current values are presented in these two forms. The general trend of the viscosity is decreasing with the increase of the temperature since this is expected for all the types of oils. For the 0 % blend of waste cooking oil (fully synthetic), there is about 900 % drop in the viscosity when the temperature increased from 10 to 80 °C.

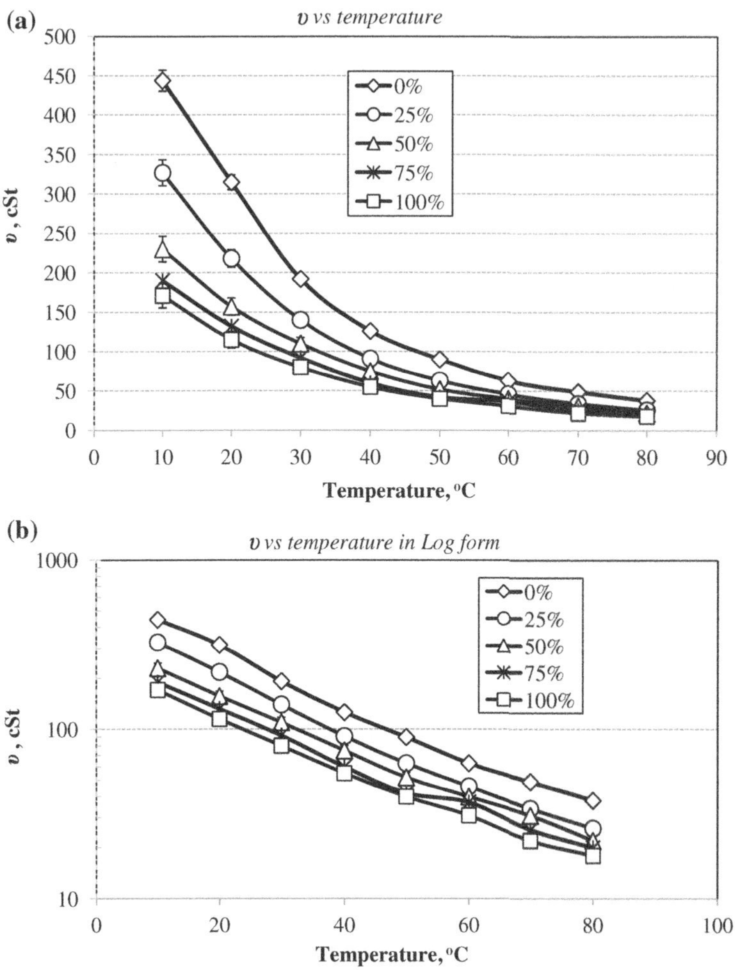

Fig. 2 Viscosity versus temperature of different blends of waste cooking and synthetic oils

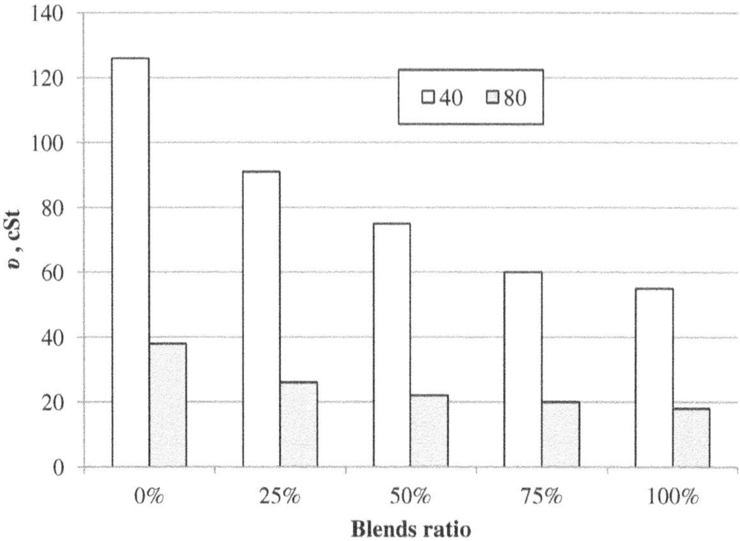

Fig. 3 Viscosity of the blends at 40 and 80 °C

Meanwhile, the drop in the viscosity of the pure waste cooking oil is about 600 %. This introduces promising results for waste cooking oil in term of stability and less sensitivity to the temperature compare to the synthetic fibers. Similar trend can be noticed in Fig. 2b.

It is well known that the determination of the oil application and performance is based on the viscosities at 40 and 80 °C. Therefore, the viscosities of the blends at those temperatures are extracted from Fig. 2 and represented in Fig. 3. It is obvious that the synthetic oil has higher viscosity values compared to its blends. Moreover, the increase in the addition of the waste cooking oil drops the viscosity of the blends for both selected temperatures as can be seen in Fig. 3.

From [75–77], the ISO viscosity grade requirements are listed in Table 4. For the current blends, it can be seen that the pure waste cooking can fit with the ISo VG 68 which can be used for crankcase oil Grades 20 W [78, 79]. However, further study is required to determine. However, further study is required to determine the degradability of the oil and the degradability of the oil and the tribological performance of the components under this lubrication condition.

Table 4 ISO viscosity grade requirement extracted from [75–77]

Kinematic viscosity	ISO VG32	ISO VG46	ISO VG68	ISO VG100	Pure waste cooking oil
@40 °C	>28.8	>41.4	>61.4	>90	65.5
@100 °C	>4.1	>4.1	>4.1	>4.1	9.5

Table 5 Compared with previous works and standards

	υ (cSt) at 40 °C	υ (cSt) 80 ° C
0	126	38
25	94	26
50	75	22
75	60	20
100	55	18
Original soybean [12]	≈175	≈29
Jatropha oil [81]	≈172	≈23
Soybean oil (SYO) [36, 45]	33.6 ± 0.9	12.3 ± 0.5
Sunflower oil (SO) [36, 45]	32.9 ± 2.3	12.7 ± 0.8
Castor oil (CO) [36, 45]	242.5 ± 21.7	37.1 ± 1.7
Castor seeds [55]	248.8	NA
Diluent (polyalphaolefin), and high-oleic vegetable oils [22]	42.33	
BIO-H01 is a mixture of 83.5 % high-oleic sunflower oil and 13.5 % ditridecyl adipate [80]	37.41	11.42
BIO-H02 blend of high-oleic sunflower oil at 73 and 24 % of diisooctyl adipate [80]	27.67	9.08

From the literature, some potential vegetable oils have been investigated to find the possibility of using them as lubricant. The most recent works are summarized in Table 5 showing the viscosity of vegetable oils at 40 and 80 °C. One can see that there are different ranges of viscosities for the oil and there is no pronounce comparable can be drawn. However, there could be comparable values for the pure waste cooking oil with soybean, [36, 45] and sunflower oil [36, 45], despite waste cooking oil exhibits better value of viscosity. In some work [80], modifier can be used to improve the viscosity performance of vegetable oil and this can be considered in the future work for waste cooking oil.

5 Conclusions

This work covers about 80 international articles published in the area of biolubricant aiming to address the most recent issues and explore the potential of suing waste cooking oil lubricant. Some important points can be drawn as conclusion from this work as follows:

1. There is a concern from environmental point of view to find alternative lubricants and an attention is paid by the researchers and vast work is focusing on the

possibility of using vegetable oils. It is highly recommended to use nonedible oils rather than edible oils.

2. There is an issue with disposing waste cooking oil. The potential of using it has been explored to be used as alternative fuel for diesel engines. However, there is a limitation of using the waste cooking oil as fuel since they highly impact on the engine performance from tribological point of view.
3. The potential of using waste cooking oil as alternative lubricant was investigated in the current study. From viscosity point of view, there is promising results to use such oil as lubricant. However, further study is recommended.

References

1. Pop L et al (2008) Basestock oils for lubricants from mixtures of corn oil and synthetic diesters. J Am Oil Chem Soc 85(1):71–76
2. Sadaka S, Boateng AA (2009) Pyrolysis and bio-oil. cooperative extension service. University of Arkansas, US Department of Agriculture and county governments cooperating
3. Calero J et al (2014) Development of a new biodiesel that integrates glycerol, by using CaO as heterogeneous catalyst, in the partial methanolysis of sunflower oil. Fuel 122:94–102
4. Tomic M et al (2014) Possibility of using biodiesel from sunflower oil as an additive for the improvement of lubrication properties of low-sulfur diesel fuel. Energy 65:101–108
5. Baron AM et al (2014) Transesterification of castor oil in a solvent-free medium using the lipase from Burkholderia cepacia LTEB11 immobilized on a hydrophobic support. Fuel 117 (Part A(0)):458–462
6. Silva RVS et al (2014) The analytical characterization of castor seed cake pyrolysis bio-oils by using comprehensive GC coupled to time of flight mass spectrometry. J Anal Appl Pyrol 106:152–159
7. Ofori-Boateng C, Lee KT (2013) The potential of using cocoa pod husks as green solid base catalysts for the transesterification of soybean oil into biodiesel: Effects of biodiesel on engine performance. Chem Eng J 220:395–401
8. Özener O et al (2014) Effects of soybean biodiesel on a DI diesel engine performance, emission and combustion characteristics. Fuel 115:875–883
9. Dugmore TIJ, Stark MS (2014) Effect of biodiesel on the autoxidation of lubricant base fluids. Fuel 124:91-96
10. Kouame SD, Liu E (2014) Characterization of fully and partially additized lubricant deposits by temperature programmed oxidation. Tribol Int 72:58–64
11. Omrani H et al (2014) Assessment of the oxidative stability of lubricant oil using fiber-coupled fluorescence excitation–emission matrix spectroscopy. Anal Chim Acta 811:1–12
12. Ting C-C, Chen C-C (2011) Viscosity and working efficiency analysis of soybean oil based bio-lubricants. Measurement 44(8):1337–1341
13. Li J et al (2014) Hydrolytic stability and tribological properties of N-containing heterocyclic borate esters as lubricant additives in rapeseed oil. Tribol Int 73:101–107
14. Cheenkachorn K, Fungtammasan B (2010) Development of engine oil using palm oil as a base stock for four-stroke engines. Energy 35(6):2552–2556
15. Syahrullail S et al (2011) Experimental evaluation of palm oil as lubricant in cold forward extrusion process. Int J Mech Sci 53(7):549–555
16. Jayed MH et al (2009) Environmental aspects and challenges of oilseed produced biodiesel in Southeast Asia. Renew Sustain Energy Rev 13(9):2452–2462

17. Kreivaitis R et al (2013) A comparison of pure and natural antioxidant modified rapeseed oil storage properties. Ind Crops Prod 43:511–516
18. Luo Y, Yang L, Tian M (2013) Influence of bio-lubricants on the tribological properties of Ti6Al4 V alloy. J Bionic Eng 10(1):84–89
19. Mobarak HM et al (2014) The prospects of biolubricants as alternatives in automotive applications. Renew Sustain Energy Rev 33:34–43
20. Saidur R et al (2011) A review on the performance of nanoparticles suspended with refrigerants and lubricating oils in refrigeration systems. Renew Sustain Energy Rev 15 (1):310–323
21. Wang C-C et al (2012) An overview of the effect of lubricant on the heat transfer performance on conventional refrigerants and natural refrigerant R-744. Renew Sustain Energy Rev 16 (7):5071–5086
22. Erhan SZ, Sharma BK, Perez JM (2006) Oxidation and low temperature stability of vegetable oil-based lubricants. Ind Crops Prod 24(3):292–299
23. Smith EH et al (1994) 9—Tribology, in mechanical engineers reference book 12th (edn). Butterworth-Heinemann, Oxford, pp 9/1–9/132
24. Yousif BF, El-Tayeb NSM (2008) Wear and friction characteristics of CGRP composite under wet contact condition using two different test techniques. Wear 265(5–6):856–864
25. Lawal S, Choudhury I, Nukman Y (2012) Application of vegetable oil-based metalworking fluids in machining ferrous metals—a review. Int J Mach Tools Manuf 52(1):1–12
26. Stachowiak GW, Batchelor AW (eds) (2014) Chapter 3—lubricants and their composition. In: Engineering tribology 4th edn. Butterworth-Heinemann, Boston, pp 51–104
27. Geitner FK, Bloch HP (2012) Chapter 3—machinery component failure analysis. In: Geitner FK, Bloch HP (eds) Machinery failure analysis and troubleshooting 4th edn. Butterworth-Heinemann, Oxford, pp 87–293
28. Chapman I (2014) The end of peak oil? Why this topic is still relevant despite recent denials. Energy Policy 64:93–101
29. Kim JE (2014) Energy security and climate change: how oil endowment influences alternative vehicle innovation. Energy Policy 66:400–410
30. Khondee N et al (2012) Airlift bioreactor containing chitosan-immobilized Sphingobium sp. P2 for treatment of lubricants in wastewater. J Hazard Mater 213–214:466–473
31. Panpanit S, Visvanathan C (2001) The role of bentonite addition in UF flux enhancement mechanisms for oil/water emulsion. J Membr Sci 184(1):59–68
32. Yusaf TF, Yousif BF, Elawad MM (2011) Crude palm oil fuel for diesel-engines: Experimental and ANN simulation approaches. Energy 36(8):4871–4878
33. Shashidhara Y, Jayaram S (2010) Vegetable oils as a potential cutting fluid—an evolution. Tribol Int 43(5):1073–1081
34. Li J et al (2008) The tribological chemistry of polysulfides in mineral oil and synthetic diester. Appl Surf Sci 254(22):7232–7236
35. Delgado-Zamarreño MM et al (2007) Analysis of synthetic phenolic antioxidants in edible oils by micellar electrokinetic capillary chromatography. Food Chem 100(4):1722–1727
36. Quinchia LA et al (2014) Tribological studies of potential vegetable oil-based lubricants containing environmentally friendly viscosity modifiers. Tribol Int 69:110–117
37. Atabani AE et al (2013) Non-edible vegetable oils: a critical evaluation of oil extraction, fatty acid compositions, biodiesel production, characteristics, engine performance and emissions production. Renew Sustain Energy Rev 18:211–245
38. Quinchia LA et al (2012) Low-temperature flow behaviour of vegetable oil-based lubricants. Ind Crops Prod 37(1):383–388
39. Nagendramma P, Kaul S (2012) Development of ecofriendly/biodegradable lubricants: an overview. Renew Sustain Energy Rev 16(1):764–774
40. Suarez PAZ et al (2009) Comparing the lubricity of biofuels obtained from pyrolysis and alcoholysis of soybean oil and their blends with petroleum diesel. Fuel 88(6):1143–1147
41. Shashidhara YM, Jayaram SR (2010) Vegetable oils as a potential cutting fluid—an evolution. Tribol Int 43(5–6):1073–1081

42. Lawal SA, Choudhury IA, Nukman Y (2013) A critical assessment of lubrication techniques in machining processes: a case for minimum quantity lubrication using vegetable oil-based lubricant. J Clean Prod 41:210–221
43. Yilmaz N (2011) Temperature-dependent viscosity correlations of vegetable oils and biofuel–diesel mixtures. Biomass Bioenergy 35(7):2936–2938
44. Franco Z, Nguyen QD (2011) Flow properties of vegetable oil–diesel fuel blends. Fuel 90 (2):838–843
45. Quinchia LA et al (2010) Viscosity modification of different vegetable oils with EVA copolymer for lubricant applications. Ind Crops Prod 32(3):607–612
46. Mejía JD, Salgado N, Orrego CE (2013) Effect of blends of Diesel and Palm-Castor biodiesels on viscosity, cloud point and flash point. Ind Crops Prod 43:791–797
47. Haldar SK, Ghosh BB, Nag A (2009) Studies on the comparison of performance and emission characteristics of a diesel engine using three degummed non-edible vegetable oils. Biomass Bioenergy 33(8):1013–1018
48. Rizwanul Fattah IM et al (2013) Impact of various biodiesel fuels obtained from edible and non-edible oils on engine exhaust gas and noise emissions. Renew Sustain Energy Rev 18:552–567
49. Talebian-Kiakalaieh A, Amin NAS, Mazaheri H (2013) A review on novel processes of biodiesel production from waste cooking oil. Appl Energy 104:683–710
50. Yaakob Z et al (2013) Overview of the production of biodiesel from Waste cooking oil. Renew Sustain Energy Rev 18:184–193
51. Mortier RM, Fox MF, Orszulik ST (2010) Chemistry and technology of lubricants. Springer Science + Business Media, Heidelberg
52. Chatra KS, Jayadas N, Kailas SV (2012) Natural oil-based lubricants, in green tribology. Springer, New York, pp 287–328
53. Kazmi A (2011) Advanced oil crop biorefineries. RSC Publishing
54. Syahrullail S et al (2011) Experimental evaluation of palm oil as lubricant in cold forward extrusion process. Int J Mech Sci 53(7):549–555
55. Madankar CS, Pradhan S, Naik SN (2013) Parametric study of reactive extraction of castor seed (Ricinus communis L.) for methyl ester production and its potential use as bio lubricant. Ind Crops Prod 43:283–290
56. Castro W et al (2006) A study of the oxidation and wear properties of vegetable oils: soybean oil without additives. J Am Oil Chem Soc 83(1):47–52
57. Singh AK (2011) Castor oil-based lubricant reduces smoke emission in two-stroke engines. Ind Crops Prod 33(2):287–295
58. Balasubramaniam B et al (2012) Comparative analysis for the production of fatty acid alkyl esterase using whole cell biocatalyst and purified enzyme from Rhizopus oryzae on waste cooking oil (sunflower oil). Waste Manag 32(8):1539–1547
59. Dovì VG et al (2009) Cleaner energy for sustainable future. J Clean Prod 17(10):889–895
60. Iglesias L et al (2012) A life cycle assessment comparison between centralized and decentralized biodiesel production from raw sunflower oil and waste cooking oils. J Clean Prod 37:162–171
61. Mendelsohn R, Neumann JE (2004) The impact of climate change on the United States economy. Cambridge University Press, Cambridge
62. Zhang H, Wang Q, Mortimer SR (2012) Waste cooking oil as an energy resource: review of Chinese policies. Renew Sustain Energy Rev 16(7):5225–5231
63. Pehan S et al (2009) Biodiesel influence on tribology characteristics of a diesel engine. Fuel 88 (6):970–979
64. Galle J et al (2012) Failure of fuel injectors in a medium speed diesel engine operating on bio-oil. Biomass Bioenergy 40:27–35
65. Demirbas A (2006) Biodiesel production via non-catalytic SCF method and biodiesel fuel characteristics. Energy Convers Manag 47(15–16):2271–2282
66. Giannelos PN et al (2005) Physical, chemical and fuel related properties of tomato seed oil for evaluating its direct use in diesel engines. Ind Crops Prod 22(3):193–199

67. Kim SW et al (2013) Bio-oil from the pyrolysis of palm and Jatropha wastes in a fluidized bed. Fuel Process Technol 108:118–124
68. Abdul Aziz, SM et al (2013) Bio-oils from microwave pyrolysis of agricultural wastes. Fuel Proc Technol 106:744–750
69. Muralidharan K, Vasudevan D, Sheeba KN (2011) Performance, emission and combustion characteristics of biodiesel fuelled variable compression ratio engine. Energy 36(8):5385–5393
70. Patil P et al (2010) Conversion of waste cooking oil to biodiesel using ferric sulfate and supercritical methanol processes. Fuel 89(2):360–364
71. Sharma BK, Adhvaryu A, Erhan SZ (2009) Friction and wear behavior of thioether hydroxy vegetable oil. Tribol Int 42(2):353–358
72. Kumar M, Bijwe J (2011) Composite friction materials based on metallic fillers: sensitivity of μ to operating variables. Tribol Int 44(2):106–113
73. Prasad BK (2011) Sliding wear response of a grey cast iron: effects of some experimental parameters. Tribol Int 44(5):660–667
74. Tewari A (2012) Load dependence of oxidative wear in metal/ceramic tribocouples in fretting environment. Wear 289:95–103
75. Mixed-fiber diets and cholesterol metabolism in middle-aged men (1991) Nutr Rev 49 (3):80–82
76. Rudnick L (2006) Automotives gear lubricants, synthetics, mineral oils, and bio-based lubricants: chemistry and technology. Taylor and Francis, Florida
77. Rudnick LR (2010) Lubricant additives: chemistry and applications. CRC Press, Florida
78. Brockwell K, Dmochowski W, Decamillo S (2004) An investigation of the steady-state performance of a pivoted shoe journal bearing with ISO VG 32 and VG 68 oils. Tribol Trans 47(4):480–488
79. Stachowiak G, Batchelor AW (2013) Engineering tribology. Butterworth-Heinemann, Oxford
80. Paredes X et al (2014) High pressure viscosity characterization of four vegetable and mineral hydraulic oils. Ind Crops Prod 54:281–290
81. Shahabuddin M et al (2013) Comparative tribological investigation of bio-lubricant formulated from a non-edible oil source (Jatropha oil). Ind Crops Prod 47:323–330

Chapter 6
Two-Body Abrasion of Bamboo Fibre/Epoxy Composites

A. Oun and B.F. Yousif

Abstract In this chapter, a new polymeric composite was developed. Bamboo fibre reinforced epoxy (BFRE) was build using unidirectional configuration and hand lay-up techniques. The bamboo fibres were treated with 6 % NaOH to improve their interfacial adhesion as recommended by the literature. The tribological performance of the new composites was investigated in terms of two-body abrasion considering different sliding distances and applied loads. Normal orientation of the fibre onto the counterface is considered in this study. For comparison purposes, neat epoxy and epoxy composites based on glass fibres were tested as well under the same conditions. Different grades of SiC papers were used in this study and optical microscopy assisted to observe the abrasive papers and the worn surfaces of the samples after the tests. The results revealed that the wear rate of the BFRE composites is very competitive to the glass fibre/epoxy composites and the neat epoxy as well at all the grades of the SiC papers. The grade of SiC papers significantly influenced the wear and frictional behaviour of the composites since the low grade exhibited higher wear rate and high friction coefficient due to the intimate contact of the asperities (interlocking process during the rubbing). Optical microscopy and SEM micrographs showed different damage features on the worn surfaces such as ploughing, grooving, scratching which represented the abrasive wear. SiC papers were either covered by patches of the soft part (epoxy or its composites) or contained loss debris from the soft parts.

1 Introduction

Nowadays, the increased demand for the use of environmentally friendly composite materials based on natural fibres is an attractive idea to many researchers in the area of tribology. This is basically due to the advantages of these fibres compared to

A. Oun · B.F. Yousif (✉)
Faculty of Health, Engineering and Sciences, University of Southern Queensland, Toowoomba, QLD 4350, Australia
e-mail: Belal.Yousif@usq.edu.au

J.P. Davim (ed.), *Ecotribology*, Materials Forming, Machining and Tribology,
DOI 10.1007/978-3-319-24007-7_6

synthetic fibres which are low cost, non-abrasive, recyclable, and possess good mechanical properties. Therefore, many studies focus on the opportunity of using natural fibres in the reinforcement of polymer composites instead of inorganic fibres. Bamboo fibre is considered to be one of these natural fibres, having a small diameter, and strong interfacial adhesion to resin matrix. Recently, many investigations have been conducted on the tribological properties of polymer composites reinforced with natural fibres such as kenaf, coir, oil palm, and jute fibres. The outcomes of these works have revealed a very high coefficient of friction in the polymer composites and low wear performance because of poor interfacial bonding between those fibres and the polyester which leads to a removal of the high layer of composite surface. On the other hand, the frictional and wear characteristics of the fibres reinforced epoxy composites were controlled using chemical treatments like alkali treatment in order to improve the interfacial bonding among the fibres and the matrix. For example, Chand and Dwivedi [1] have reported improving abrasive wear resistance of the treated jute fibres with the epoxy matrix compared to untreated fibres. With regards to the reinforcement of polymer composites by using natural fibres as filler for tribological application, few works have been carried out to investigate the tribological performance of natural fibres in abrasive and adhesive wear modes [11, 2]. The current research examines the adhesive wear and two-body abrasion of bamboo fibre/polyester composites. Three materials were selected for this research and were tested under different dry contact conditions (applied load and sliding distance) using different abrasive paper grades. The literature reveals studies into the tribology behaviour of polyester composites reinforced with natural fibres like betel nut [18], kenaf, coir, sugarcane [7], jute [1]. Although these studies presented various surface damage features which resulted from tests at different sliding conditions, not one of them focused on the adhesive wear and two-body abrasion of the polyester composites. The main aim of this project is to develop a new biocomposite and test its tribological performance.

2 Material Preparation and Experimental Procedure

2.1 Material Preparation

The materials used in this project were the same as materials that have been used in the previous studies by many researchers such as [8, 11, 12]. They are bamboo fibre reinforced epoxy (BFRE), glass fibre reinforced epoxy (GFRE), and neat epoxy (NE). The objective of the use of these materials is to show the characteristics of surface damage when tests are conducted under different dry loading conditions. So the possibility of attempt with new materials may not clearly show surface damage as in these materials. The experiments were conducted in three different abrasive paper grades which are G1200, G400, and G80.

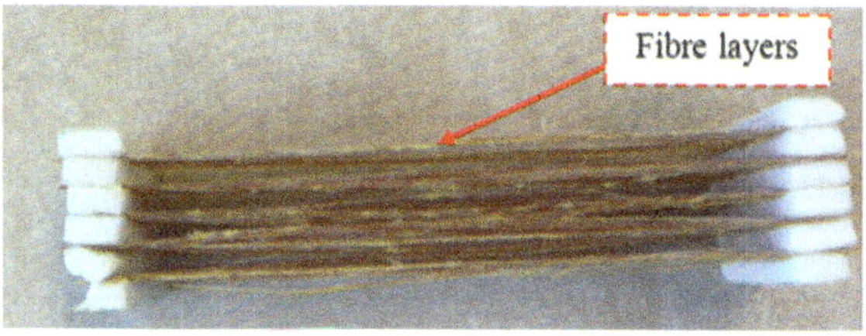

Fig. 1 Picture of prepared bamboo fibre structure

Raw bamboo fibres were provided from the Malaysian Agricultural Research and Development Institute (MARDI). The fibres were sorted in equal diameters and then the process of combing these fibres was conducted before placing them into a plastic mould with 6 % of NaOH solution. The fibres were left immersed in chemical treatment for 24 h at a room temperature of 27 °C. After the treating process, the fibre surface was thoroughly cleaned of impurities and increased roughness of fibre surface was clearly observed. This was due to the alkaline treatment which leads to the possibility of enhancing the interfacial bond strength between the fibre and the epoxy matrix as reported by Yousif and El-Tayeb [17]. Also, according to Nirmal et al. [10], the chemical treatment helps to enhance the wear performance and reduces the porosity of a composite. In the next step, the bamboo fibres were cut into a uniform length of 170 mm and placed in a unidirectional format. Before being placed into a metal mould, the end of fibre sides were pasted together with double-sided tape to make the first layer of fibre structure. This step was repeated until the completion of the composite structure as shown in Fig. 1.

2.2 *Preparation of Composites*

The fabrication process of the composite has begun by coating the inner walls of the metal mould with a light layer of wax which works to prevent sticking between the mixture and the mould allowing easy removal after the curing process. Treated bamboo fibres were organized and placed into a metal mould, with dimensions 70 mm × 10 mm × 10 mm, in a unidirectional form. In this work, a liquid of epoxy resin (DER331) is used. The ratio of epoxy resin to hardener in this study is 2:1. The mixture was uniformly mixed as a result of using an electric stir and was then poured into the metal mould. To avoid generating air bubbles, the mixture sat between 5 to 10 min before being pouring into the metal mould. The mould remained at a vacuum room (MCP004PLC) at 27 °C to cure for 24 h. After that the mixture was treated in an oven for another 24 h at 80 °C with 10 kPa pressure. Fibre volume fraction in the matrix was around 48 % Vol. Bamboo fibre reinforced epoxy (BFRE), glass fibre reinforced epoxy, and neat epoxies were selected as reinforcement materials in this research. More information about material preparation has been given by Chin and Yousif [5]. The composite block was machined into

Fig. 2 Sample of the prepared composites

specimens of 20 mm × 10 mm × 10 mm and the tribological tests were conducted in 20 mm × 10 mm. In the fabrication process of synthetic composite, a similar method was used for mixing the epoxy liquid with hardener in the ratio of 2:1 and glass fibre replaced the bamboo fibre. The photo of the composites is presented in Fig. 2.

The sliding conditions were selected in this study depending on previous studies which have been performed to investigate tribological and mechanical behaviours of fibre polymer based on natural fibres. Of these sliding parameters for dry contact conditions, there were sliding distances (1.35–20.25 m), a sliding speed of 2.8 m/s, and changeable applied loads between 25 to 100 N. During the dry contact condition test, changeable applied loads were among 25–100 N for bamboo fibre reinforced polyester (BFRP), glass fibre reinforced polyester (GFRP), and neat epoxy (NE).

2.3 Experimental Procedure

A Block-On-Ring (BOR) machine was used to carry out the tests. This machine was adapted for wet and dry contact sliding tests. Figure 3 reveals the major components of the machine which are a stainless steel counterface, load position, load cell, specimen holder, and control panel. The load cell is used to measure frictional forces that are generated between the counterface and the specimens. During dry contact tests, the interface temperature was measured by using an infrared thermometer that was pointed in the mid-point of interface between the counterface and the specimen. With regards to the abrasive paper grade, it is fixed on the surface of the ring counterface as shown in Fig. 3.

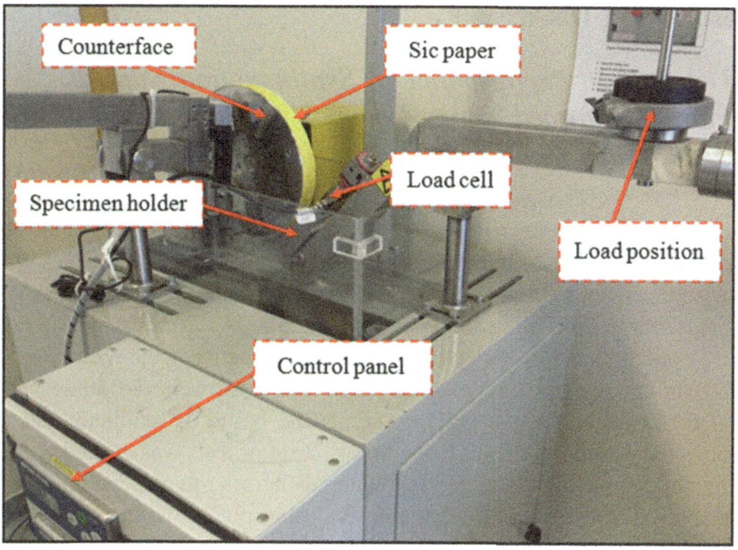

Fig. 3 Main parts of a Block-On-Ring (BOR)

Experiments were performed using a Block-On-Disk (BOD) machine. The composite surface specimens (20 mm × 10 mm × 10 mm) was tested against a firm counterface made of stainless steel (AISI 304, Ra = 0.1 μm, hardness = 1250 HD). All the specimens' contact surfaces were cleaned with a dry soft brush. The counterface surface was cleaned with a wet cloth and acetone, and then dried by a soft cloth and hot air at around 100 °C for 5 min before being fixed the abrasive paper grade.

To ensure more intimate contact between the counterface and abrasive paper, they were pasted by double-sided tape. Moreover, the specimen was fixed in the specimen holder to ensure that it did not remove during the test ensuring full contact with the counterface. The experiments were performed in changeable applied loads between 25 and 100 N, sliding speed of 2.8 m/s, and sliding distances (1.35–20.25 m) in room temperature of 27 °C. The dry soft brush was used to clean the prepared composite samples pre and post the test. Prior to and after the test, weights of all specimens were determined by using Setra weight balancer (±0.1 mg) and then weight loss was computed. Friction force was determined via load cell during the tests.

The SEM machine was used to study the composite surface morphology. The specimen surfaces were covered with a light layer of platinum before using the SEM machine. Before tabulating the average results, every tribological test was repeated three times. Many various surface observation methods can be used to show surface damage and to obtain the required results, including optical microscopy and Joel SEM machine.

3 Results and Discussion

3.1 *Tribological Performance of Synthetic and Natural Fibre Reinforced Epoxy Composites Under Different Dry Contact Conditions*

The experimental outcomes of the neat epoxy, glass, and BFRE composites are presented in this chapter under different operating parameters using different abrasive paper grades. Surface observations are introduced in the wear rate of the composite surface, and the coefficient of friction by using scanning electron microscopy and optical microscopy. The relationship between the wear rate, applied load, and sliding distance are plotted in the charts as well. To study wear and frictional performance for the polymer composite based on synthetic and natural fibres subjected to different sliding conditions, a set of tests was carried on the neat epoxy, bamboo/epoxy, and glass/epoxy materials in order to achieve this aim under different applied loads using different abrasive paper grades, with a sliding distance from 1.35 to 20.25 m. The results are represented on individual charts. The tribological outcomes of the polymeric composite reinforced with neat epoxy, glass epoxy composite, and bamboo epoxy composite are plotted in Figs. 4 to 14 with comparisons between them. The graphs illustrate the wear rate and the friction coefficient of the selected materials under different applied loads using different abrasive paper grades.

3.2 *Wear Performance of Bamboo*

Figure 4 displays the wear rate of the bamboo epoxy composite under different applied loads using different abrasive paper grades. Figure 4a demonstrates an increase in the wear rate values of the BFRE composite in the lower range of the applied load between 25 and 50 N. An increase in the applied load of more than 50 N has a significant impact on the values of the wear rate. In Fig. 4b, c shows similar findings compared to the previous figure when the applied load increases. The increase in the applied load showed reduction in the wear rate of bamboo epoxy composite and this is mainly due to the end of the fibres helping to carry the load out of the polyester region and resisting the shear strength. Further discussion is given in the scanning electron microscopy section.

There is a decrease in the wear rate values of bamboo fibre reinforced epoxy composite when the applied load increases and this result is presented in Fig. 4a–c. All of these charts show the steady state of the wear rate of the BFRE after around 15 m sliding distance, but there is a very high wear rate recorded before this sliding distance. It can be observed that all figures display an increase in the wear rate value

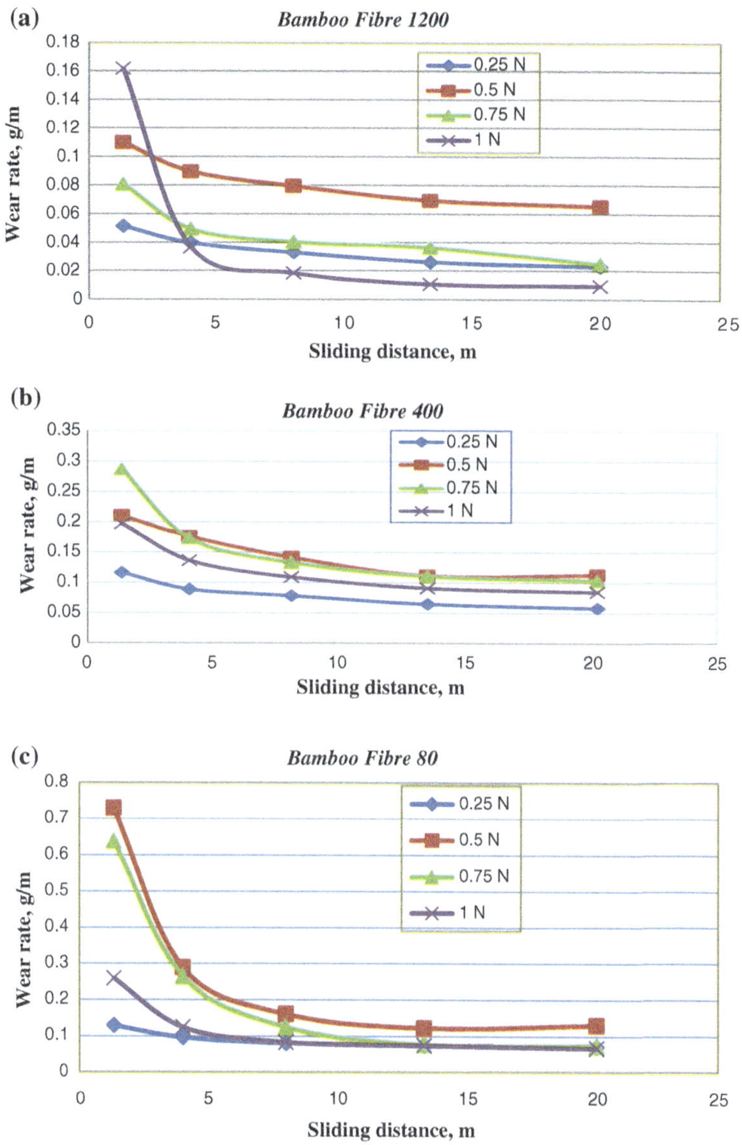

Fig. 4 Wear rate of bamboo fibre reinforced epoxy composite under different applied loads using different abrasive paper grades

of the bamboo epoxy composite in the lower range of the applied load between 25 and 50 N. The highest wear rate value is recorded at applied load (50 N), while the lowest value of wear rate is recorded with an increasing load applied to 75 N and then to 1 N compared to the previous value of the applied load (50 N). This means that there is the possibility of replacing the glass fibre with bamboo fibre due to decreased wear rate. The major reason for the decrease in wear rate is the high interfacial adhesion of bamboo fibres to the synthetic matrix. This has been investigated by some recent researchers such as [4, 14]. Generally speaking, all tested parameters (such as applied load and sliding distance) have an effect on the wear outcomes.

3.2.1 Neat Epoxy Wear Rate

Figure 5 shows the wear rate of the neat epoxy under different applied loads using different abrasive paper grades. Figure 5a shows that the increase in the sliding distance decreases the wear rate for all the applied loads. For the applied load, there is no clear trend can be drawn. Figure 5b shows similar results as in the previous Fig. 4a when the neat epoxy is subjected to the increase in applied load, however, at the grade of 80 Fig. 5c shows that the increased applied load increases the wear rate.

The increase in the sliding distance showed a decrease in the wear rate and this could be due to the filling of the abrasive paper while the rubbing process continues. In other words, a steady state is reached after about 15 m. This will be further explained with the aid of the optical microscopy in the next sections. From the findings shown in Fig. 5, it can be seen that a high degree of outcome fluctuation at the outcomes could be observed at all applied loads. This fluctuation phenomenon is due to the clean surface of abrasive paper in the beginning run period when the wear is initiated in epoxy region [17, 14, 19]. Wear rates begin a gradual decline with increasing sliding distance until a steady state condition is reached.

3.2.2 Wear Performance of Glass

Figure 6 presents the wear rate of the glass epoxy composite under different applied loads using different abrasive paper grades. Figure 6 demonstrates an increase in the wear rate values of GFRE in response to the increase in sliding distance. In Fig. 6b, a similar trend of an increasing wear rate value was observed with an increase in sliding distance for all applied loads, however, in the grade of 400 Fig. 6c displays that the wear rate curve has no clear trend at a sliding distance between 1.35 and 20.25 m for all applied loads. In general, the increase in the sliding distance showed increase in the wear rate and this is due to the brittleness of glass fibres. This will be further clarified in the surface observation section.

From the test results obtained, the charts explain that after a 15 m sliding distance (steady state condition), the value of the wear rate of the glass epoxy

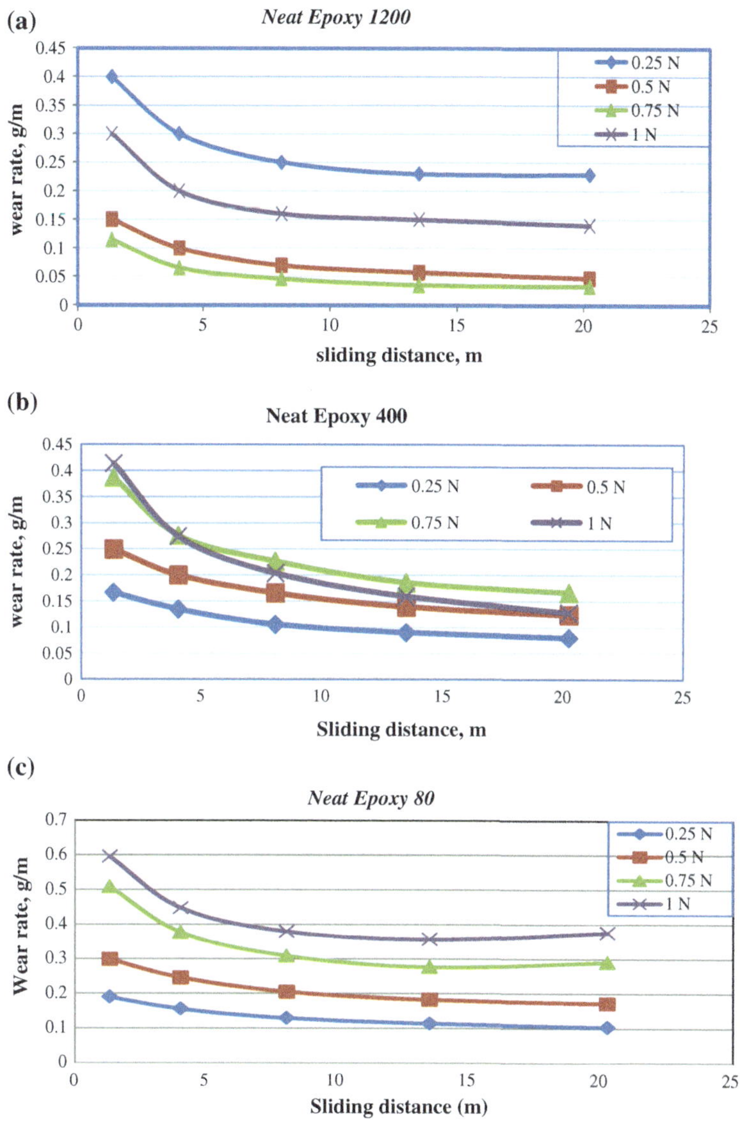

Fig. 5 Wear rate of neat epoxy under different applied loads using different abrasive paper grades

composite increased and was in the range of 0.1–0.4 g/m for all applied loads. The average of wear rate was initially between 0.1 and 0.2 g/m for a sliding distance between 1.35 and 20.25 m. However, the wear rate increased when the applied load increased. Also, a similar trend in increasing wear rate was observed with other

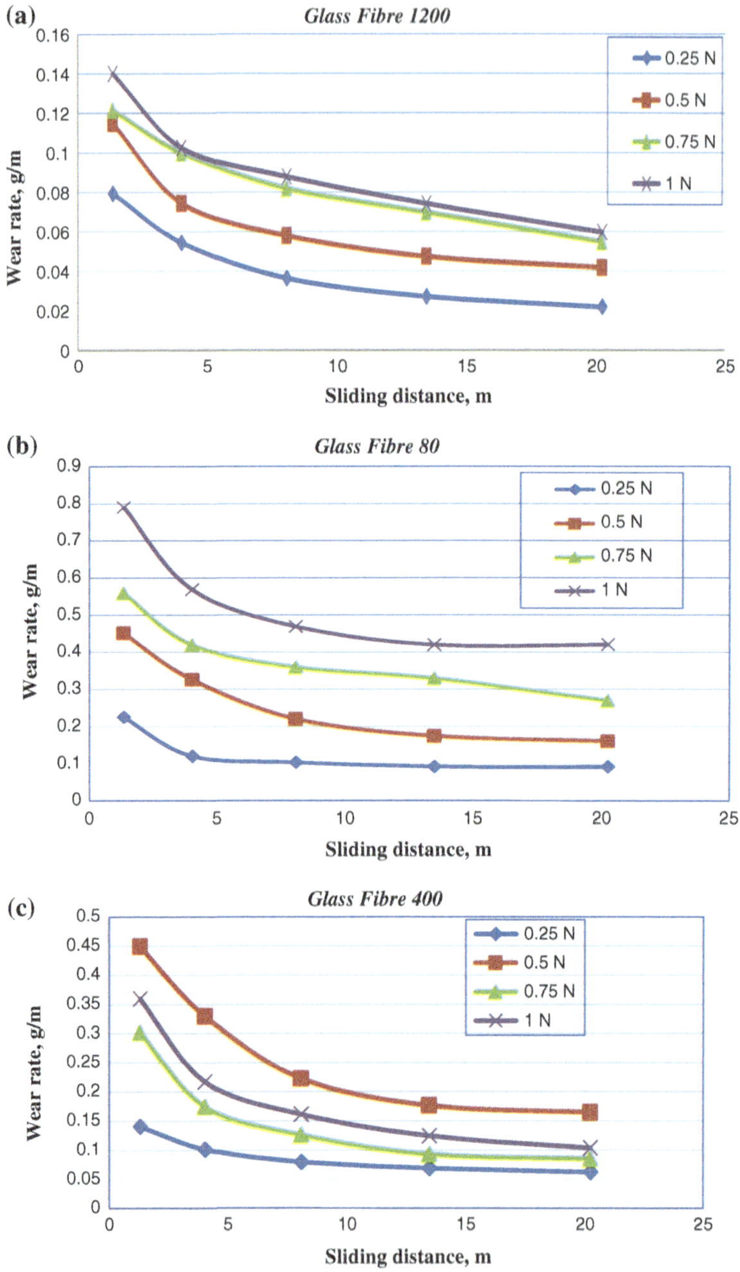

Fig. 6 Wear rate of glass fibre reinforced epoxy composite under different applied loads using different abrasive paper grades

synthetic fibre reinforced polyester composites. Many works have reported this phenomenon, such as [10, 17]. Overall, the curve of the wear rate of the glass epoxy composite has an increasing trend in all applied loads with respect to sliding distance and this suggests another potential for replacing the synthetic fibre with bamboo fibre.

3.2.3 Comparison of Wear Performance of Neat Epoxy and Its Composites

The comparison between the wear performances of the BFRE composite, the GFRE composite and neat epoxy (NE) under an applied load of 1 N using different abrasive paper grades (presented in Fig. 7a–c) shows the remarkable influence of operating parameters on the wear outcomes. All graphs clearly display that at a sliding distance of 20 m, the wear rates of the BFRE composite for all grades have lower values compared to others. This indicates that the wear rate values of the bamboo epoxy composite for all applied loads decrease when the sliding distance increases. At the higher applied load of 1 N, the high values of the wear rate in the neat epoxy followed the GFRE composite at all grades except for the grade of 80, the maximum value of the wear rate in the GFRE composite followed the neat epoxy. Steady state conditions for all materials were reached after about 15 m sliding distance. From the above outcomes, the bamboo epoxy composite demonstrates higher wear resistance in all applied loads compared to other fibres and this is due to stronger interfacial adhesion of bamboo fibres with the epoxy matrix, which prevents the bending and pulls out of the fibres at the rubbing area. At sliding distance of 20 m, the average results for the wear rate of BFRE were between 0.02 and 0.12 g/m, while for the glass epoxy composite results were in the range of 0.06–0.4 g/m, and the neat epoxy was between 0.14 and 0.39 g/m at applied load of 1 N. In other words, there is a remarkable difference in the wear rate for the neat epoxy and its composites under the same applied load and siding distance of 20 m. From the figures below, it can be noted that the presence of bamboo fibres in the polyester matrix improved the wear behaviour of the neat epoxy. Furthermore, the graphs indicate that the composite based on natural fibre possesses better wear behaviour than the synthetic fibre. In other words, bamboo fibres help in the reduction of the wear rate of the epoxy particularly when the applied load increases. This means that for adhesive wear applications there is the potential of using bamboo fibre to reinforce polyester composites and replacing glass fibres. This is due to the promising results of bamboo fibres which have a high interfacial adhesion between the fibre and the epoxy matrix. This has been reported by many researchers like [10, 17].

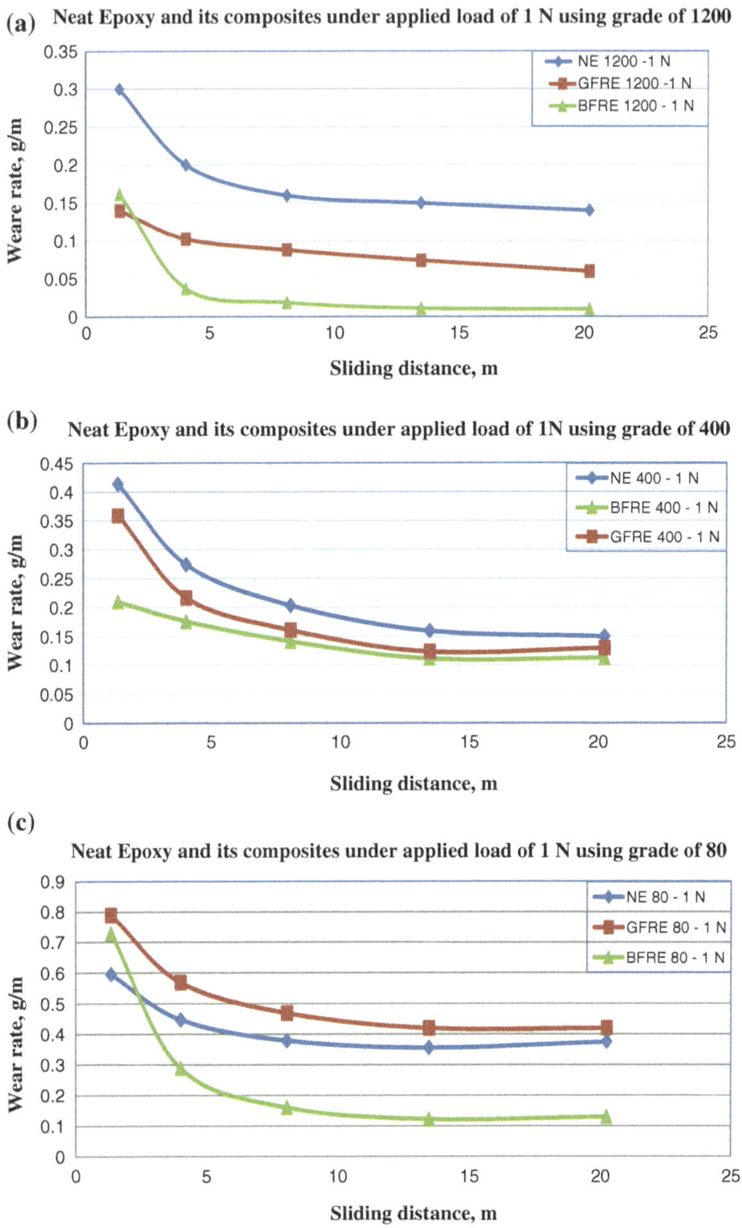

Fig. 7 Wear performances for neat epoxy and its composites under applied load of 1 N using different abrasive paper grades

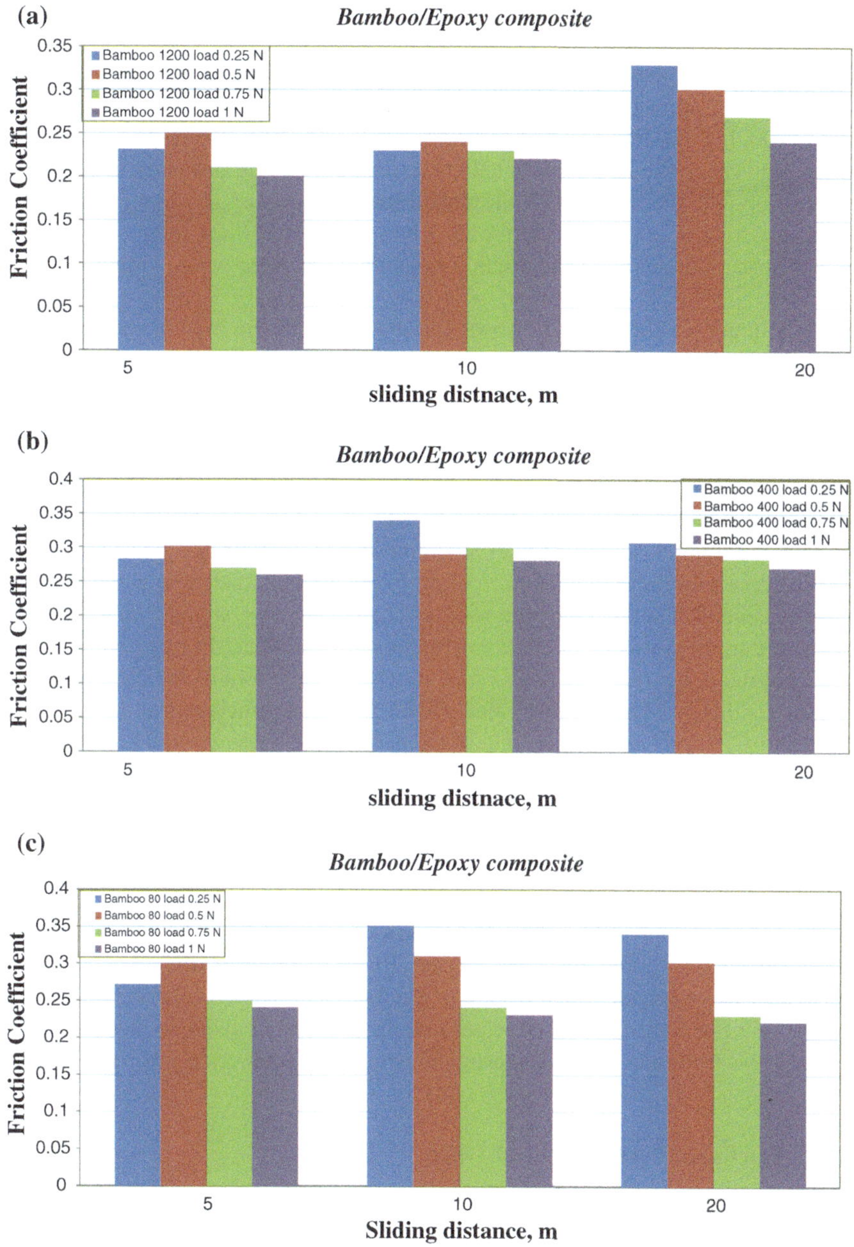

Fig. 8 Frictional performance of bamboo fibre reinforced epoxy composite under different applied loads using different abrasive paper grades

3.2.4 Frictional Performance of Bamboo

Figure 8 displays the friction coefficient values of the bamboo epoxy composite under different applied loads using different abrasive paper grades. In the case of grade 1200, Fig. 8a shows that, at sliding distance of 5–10 m, the friction coefficient values are scattered and the frictional curve for both sliding distances is similar in terms of up and down values at all applied loads. However, at sliding distance of 20 m, the friction coefficient decreases when the applied load increases. For the applied load, the curve for the friction coefficient declines when sliding distance increases.

At the grade of 400, Fig. 8b shows that the values of the friction coefficient for all sliding distances are also scattered, except for sliding distance of 20 m which indicates a decrease in the friction coefficient when the applied load increases. In Fig. 8c, no clear trend for the friction coefficient can be drawn before 10 m sliding distance but, after this sliding distance, a decrease in the friction coefficient occurs when the applied load increases. The range results of the friction coefficient for the grade of 80 are between 0.23 and 0.35 where the highest value of friction coefficient was measured at 25 N applied load at sliding distance of 10 m.

The average of the coefficient of friction for all applied loads was within 0.2 and 0.35. In general, the values of the friction coefficient were slightly increased and this is due to the fibres being exposed to the sliding surface. In other words, fibre ends contribute to protect the matrix region, which leads towards the increase in the friction coefficient. The highest value of the friction coefficient was measured at applied load of 25 N at the beginning of the period which was about 0.35. While the lower value of friction coefficient can be observed at a higher applied load of 1 N which was 0.2. At higher applied loads of 75 and 1 N, the BFRE composite shows lower values of friction coefficient for all graphs and this may contribute to the improved mechanical properties of polymeric composites. Moreover, it can be observed from the graphs that the coefficient of friction did not reach a steady state condition for all applied loads. In other words, there is no clear frictional trend for all graphs at all sliding distances in the BFRE composite due to the surface adjustment by treatment which enhances the interfacial adhesion between the fibre and the polyester matrix. This indicates that the bamboo fibre epoxy composite has a good wear resistance. Many investigations have been reported this like [6, 16].

3.2.5 Frictional Performance of Glass

Figure 9 exhibits the friction coefficient values of the glass epoxy composite under different applied loads using different abrasive paper grades. At grade of 1200, Fig. 9a shows that for all sliding distance increase the friction coefficient in the beginning run period and starts to be reached steady state of the friction coefficient

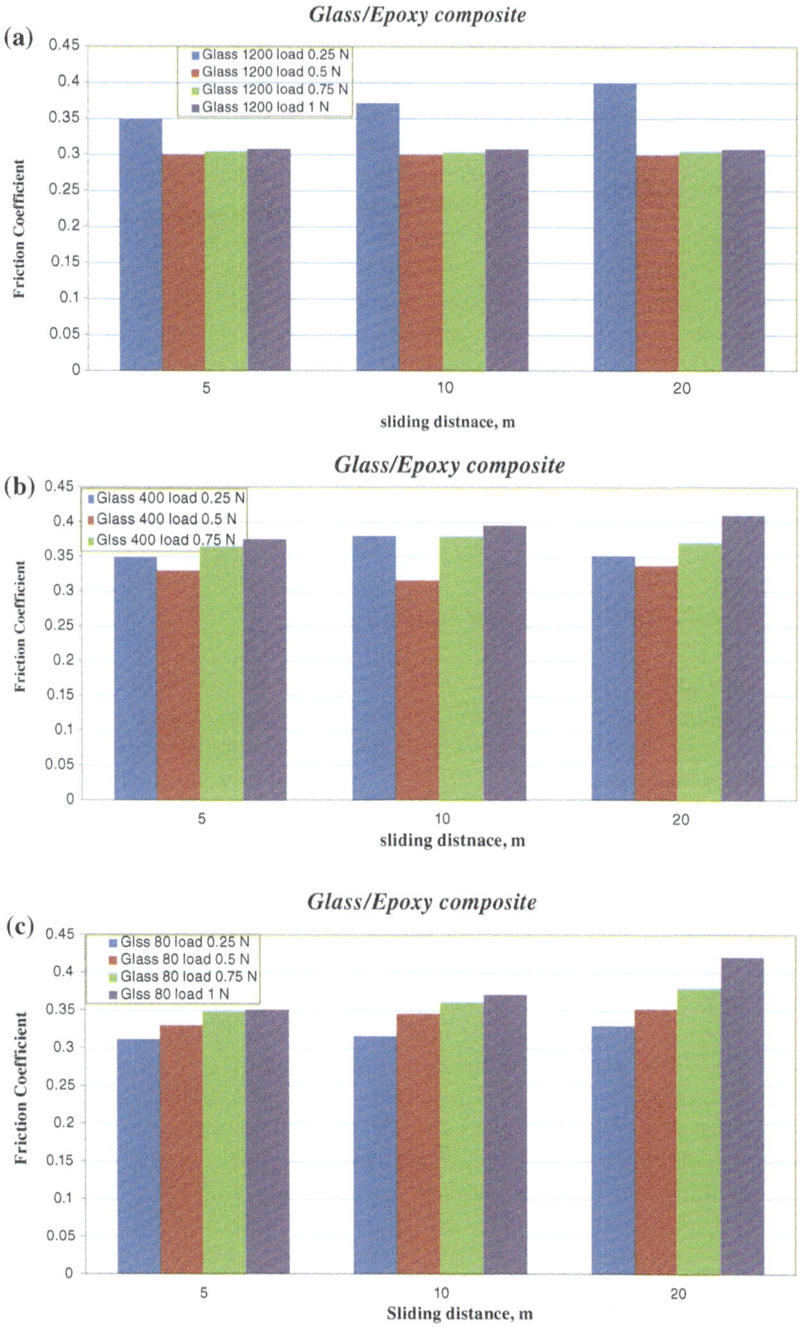

Fig. 9 Frictional performance of glass fibre reinforced epoxy composite under different applied loads using different abrasive paper grade

at different applied loads. The coefficients of friction for all applied loads are within the range of 0.29–0.31 except for the applied load of 25 N which was 0.4 at sliding distance of 20 m. Figure 9b shows the increase in the coefficient of friction more than 0.4 with the grade of 400. Moreover, it seems that the trend of the friction coefficient is not clear at all sliding distances. For the grade of 80, Fig. 9c shows that, for all sliding distances, the increase in the applied load increases the friction coefficient.

At different applied loads, the glass fibre demonstrates a friction coefficient above 0.25–0.42. In general, the values of friction coefficient of glass are higher compared to the bamboo. The lower friction coefficient of glass fibre can be achieved with the grade of 1200 at applied load of 50 N and sliding distance of 10, while the high friction coefficient of the composite can be observed in the grade of 80 with the applied load of 1 followed by the grade of 400. For all grades, there is an impact of sliding distance on the frictional behaviour of the composite at higher applied loads, especially at the grade of 80 where the increase in the applied load has a significant influence on the values of the friction coefficient. Each curve displayed a different trend in frictional forces as a function of sliding distance.

3.2.6 Frictional Performance of Neat Epoxy

Figure 10 shows the friction coefficient values of the neat epoxy composite under different applied loads using different abrasive paper grades. Figure 10a shows that the neat epoxy for all sliding distances demonstrates higher values of friction coefficient at an applied load of 25 N. The average outcomes for friction coefficient of the neat epoxy are between 0.31 and 0.42 at the grade of 1200 at all applied loads. At grade of 400, Fig. 10b shows that the increase in the applied load increases the coefficient of friction at a 5 m sliding distance, while for other sliding distances the values of the coefficient of friction are scattered due to the high contact pressure resulting from the various applied loads on the neat epoxy, which can cause limited bonding between hardener and epoxy resin. As a result of this, the inability of the neat epoxy to resist lengthy exposure for sliding distances can lead to micro- and macro cracks on the neat epoxy surface. In this chart, the highest friction value was measured at the high applied load of 1 N which is 0.39 at a sliding distance of 5 m, while the lowest value was 0.3 at a sliding distance of 10 m. For the grade of 80, Fig. 10c shows a similar trend in the friction coefficient to the previous chart Fig. 10b. In other words, the neat epoxy with the grade of 80 confirms similar friction performance to that of the grade of 400. This means that the applied load does not influence frictional performance of neat epoxy using different abrasive paper grades. Generally, the friction coefficient values are high, and are considered to be one of many issues that the adhesive wear performance of neat polyesters has. This has been reported by many authors like Suresha et al. [15]

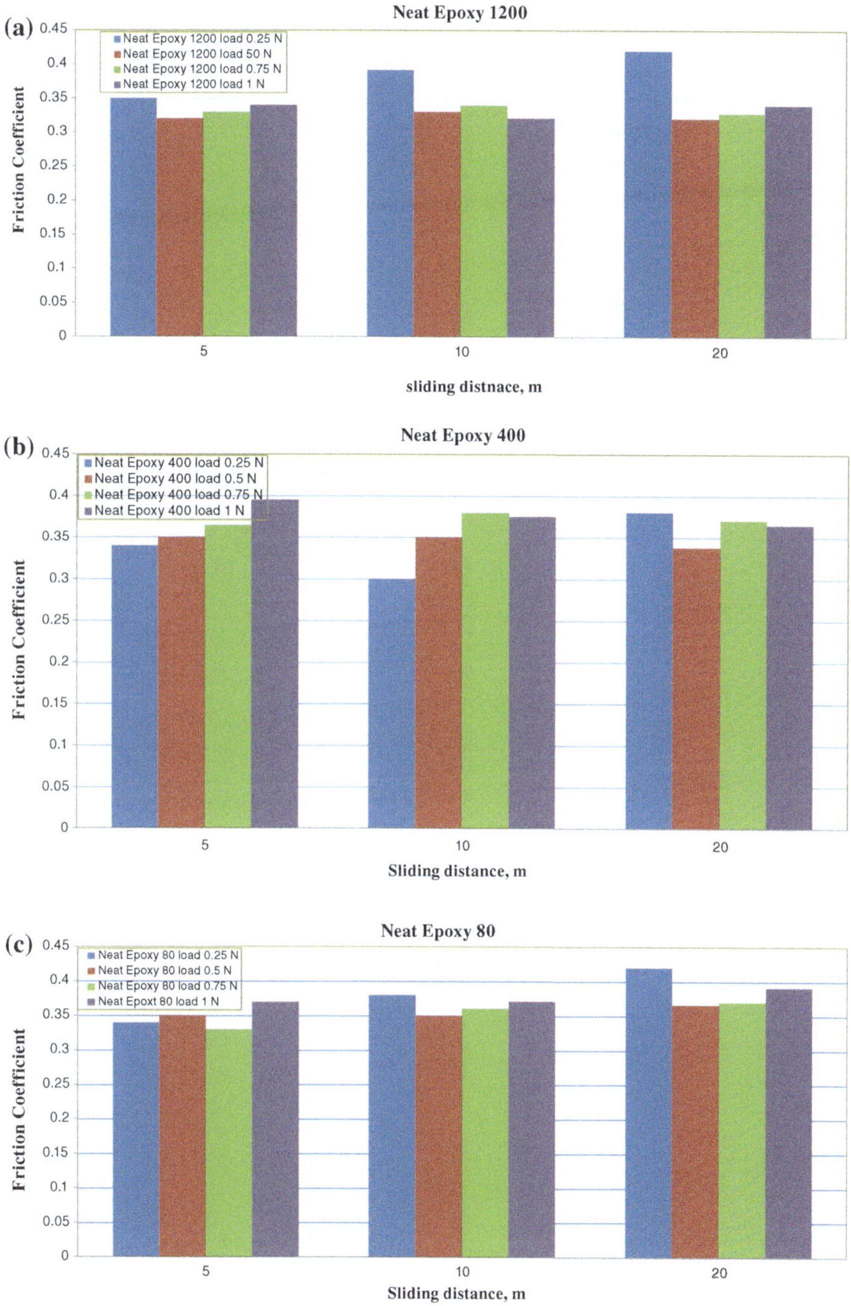

Fig. 10 Frictional performance of neat epoxy under different applied loads using different abrasive paper grades

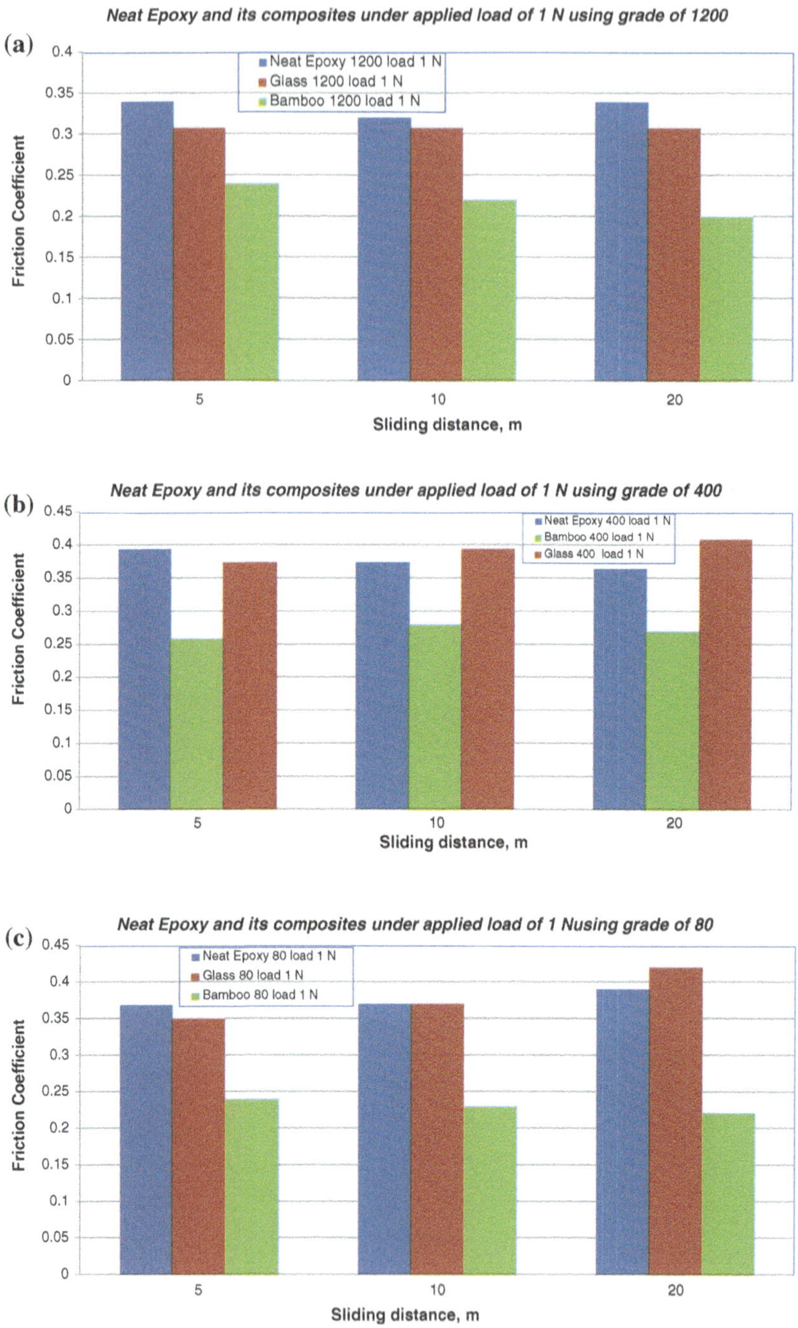

Fig. 11 Frictional performances for neat epoxy and its composites under applied load of 1 N using different abrasive paper grades

and Sharma et al. [13]. The results of the friction coefficient for all graphs show close values at all applied loads which is due to homogeneous asperities at contact. All charts show that the high values of the friction coefficient led to an increase in the wear rate of the neat epoxy in the rubbing area, which could be due to the increase in the interface temperature on the neat epoxy surface. It is believed that the presence of natural fibres such bamboo fibre on the neat epoxy surface might achieve better frictional behaviour as has been studied in recent research such as [17, 18].

3.2.7 Comparison of Frictional Performance of Neat Epoxy and Its Composites

Compared with the frictional performances of the BFRE composite, the GFRE composite and neat epoxy (NE) under an applied load of 1 N using different abrasive paper grades presented in Fig. 11a–c, each curve distribution demonstrates a different trend of friction coefficients. Overall, it can be seen that polyester composites reinforced with bamboo introduce better frictional behaviour than the others. The graphs demonstrate that the neat epoxy gives the highest frictional values, followed by the glass fibre epoxy composite. The main reason for the occurrence of damage features (such as high material removal) on the surface of the GFRE composite and the NE is the high values of their friction coefficient which led to increasing interface temperature on their surfaces. At higher applied load, epoxy composite reinforced with glass fibre exhibits a high friction coefficient compared to epoxy composite based on bamboo fibre. This gives it the potential for replacing synthetic fibres with natural fibres to reinforce polymeric composites for tribological and mechanical applications. The charts clearly demonstrate that the frictional performance for bamboo epoxy composite is better at higher applied loads except for the grade of 1200. This indicates that the surface roughness of the bamboo epoxy composite is less than the others, which produces low-interface temperature by friction leading to low wear and friction behaviour of the composite. Although the friction coefficient for the bamboo fibres recorded slight increase in value, the values are less compared to the other fibres.

This suggests that polyester composites reinforced with bamboo fibres introduce better frictional and wear performance compared to the other natural and synthetic fibres. This is due to the high interfacial adhesion of bamboo fibres with a synthetic matrix as mentioned in the wear performance of bamboo section. Many recent researchers have reported this, including [1, 4].

3.3 Surface Observation for Neat Epoxy and Its Composites

Figure 12 displays micrographs of the worn surfaces of bamboo fibre epoxy composite under different applied loads using different abrasive paper grades.

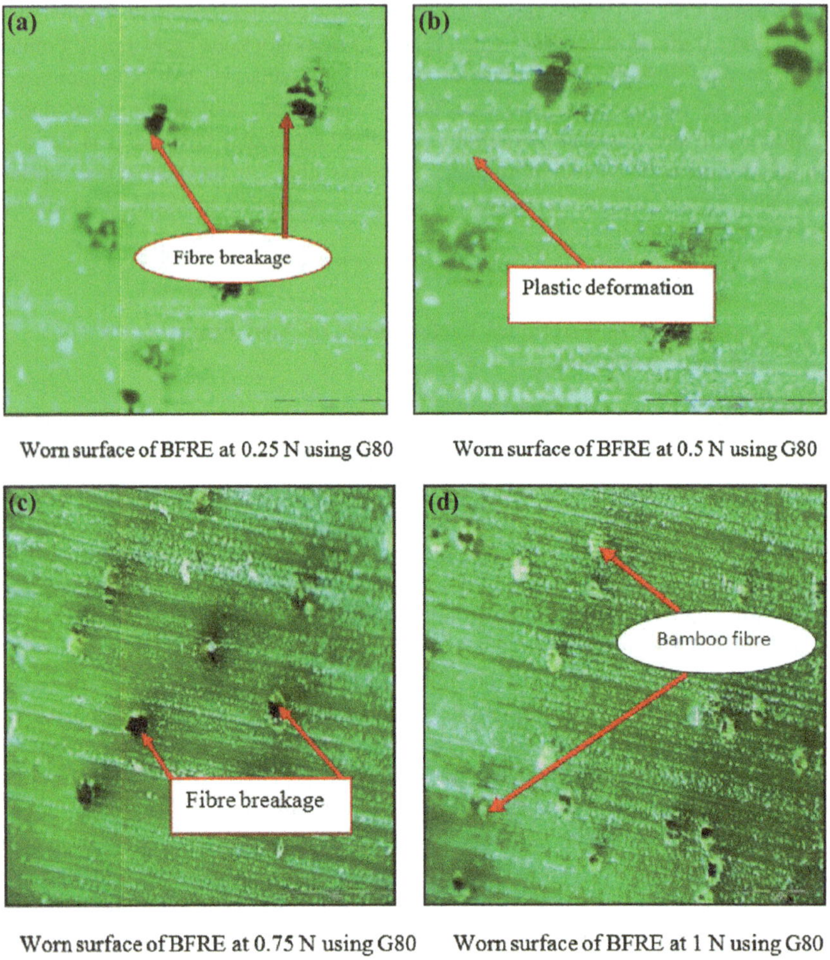

Fig. 12 Micrographs of the worn surface of the bamboo fibre reinforced epoxy under different applied loads using G80

Figure 12a shows that the softening process happened in the resinous areas and was associated with the shear loading. However, there is no trace of the pull-out of fibres on the composite surface during the rubbing process. In the case of grade of 80 and 400 (Fig. 12b, c), surface topography illustrates a clear plastic deformation on the composite surface, which is due to increased and distributed interface temperatures on the whole surface of the composite and to the increase in the surface roughness of these abrasive papers compared to the previous grade of 1200. All charts show

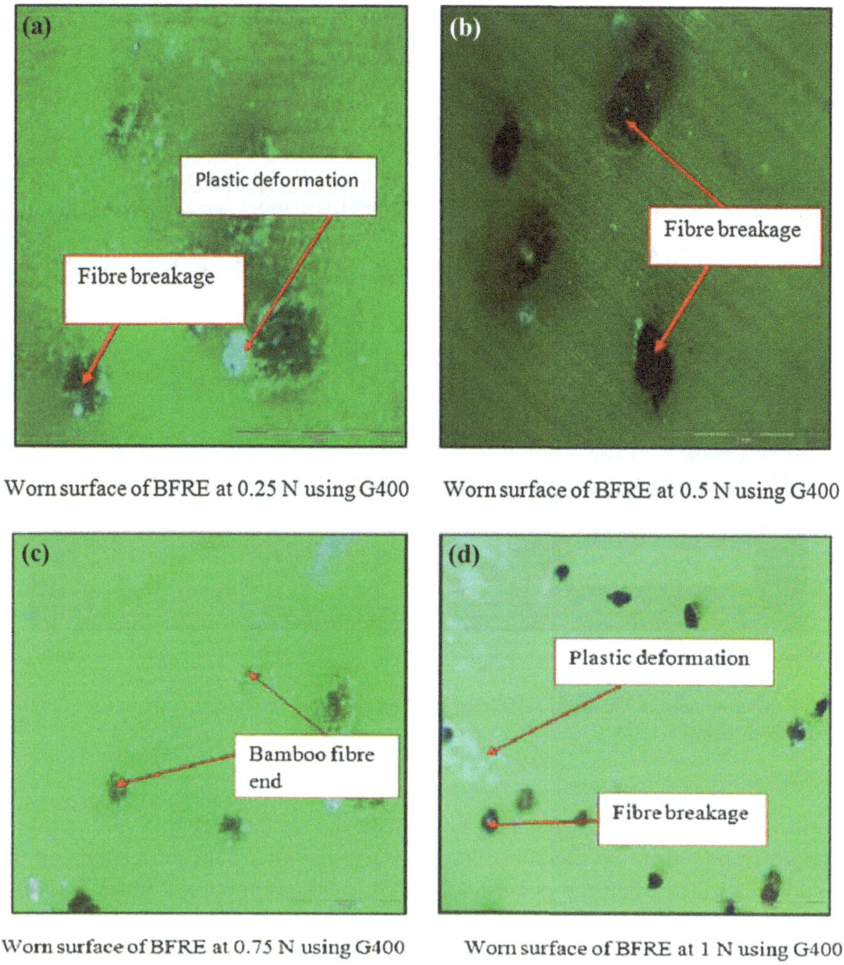

Fig. 13 Micrographs of the worn surface of the bamboo fibre reinforced epoxy under different applied loads using G400

that the bamboo fibre epoxy composite has better wear resistance. This could be due to the fibres being oriented perpendicular to the direction of the sliding surface.

In other words, the fibres have been able to resist the shear strength and carry the applied load outside the polyester area. Although peeling of fibres took place gradually which leads to detachment and breakage of fibres over time, this problem is solved by controlling the interfacial adhesion property of the fibre with the polyester matrix, as has been reported by Yousif et al. [19]. This indicates high

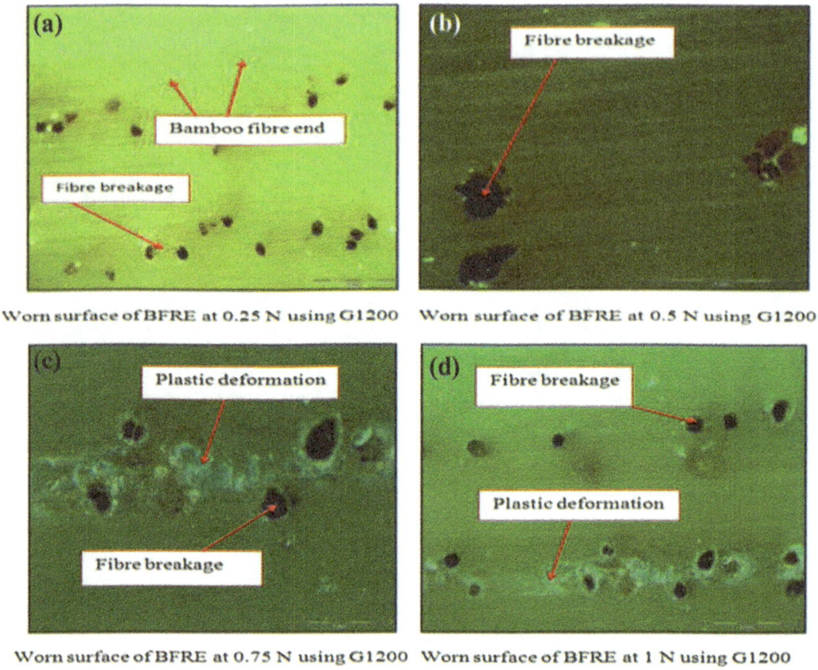

Fig. 14 Micrographs of the worn surface of the bamboo fibre reinforced epoxy under different applied loads using G1200

interfacial bond strength of bamboo fibres with the synthetic matrix. Some investigations (reported in the literature) have observed similar damage features on the composite surface when the polyester composite based on natural fibres was subjected to frictional force and heat which led to an increase in the interface temperature. They showed similar damages with kenaf [4], jute [1], and sugarcane [7]. These researchers recommend that further investigations be conducted to decrease the high frictional heat and force on the composite surface because they are considered to be the main reason for such damage. With regards to surface observation on the abrasive paper grades, the surface micrographs for all grades show that there is epoxy debris, and an epoxy patch can be seen on the abrasive paper surface as shown in Fig. 13a–c. This could be due to the high interface temperature on the composite surface which is caused by friction.

The micrographs of the neat epoxy surface under different applied loads using different abrasive paper grades are presented in Fig. 14a–c. All charts show that the surface micrographs of neat epoxy demonstrate softening and deformation of the epoxy. This could be due to the increase in the interface temperature on the surface

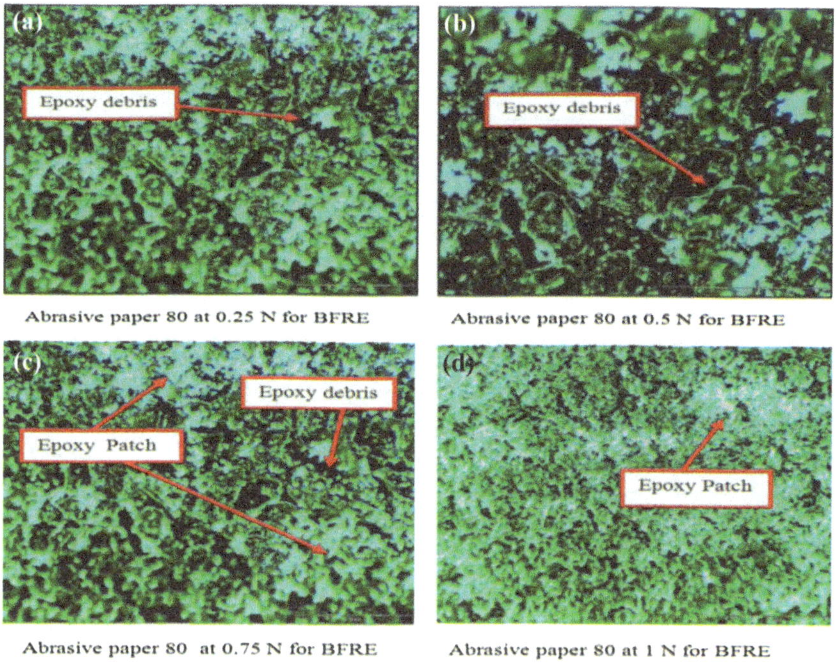

Fig. 15 Micrographs of the abrasive paper when the bamboo fibre reinforced epoxy tested under different applied loads using G80

of the epoxy when it is subjected to the applied load. This increase in interface temperature associated with the shear loading helps to create micro-cracks on the epoxy surface. At higher applied loads, the high amount of material removal can be observed in the neat epoxy. It can be also observed that areas of grooves and sharp asperities appear along travel track on the surface of the epoxy during the process of rubbing. This indicates the high weight loss of the epoxy surface which is considered one of many issues associated with neat epoxy (as mentioned at the frictional performance of neat epoxy section). Regarding the grade surface inspection, a similar observation about the neat epoxy can be seen in Fig. 15a–c.

Figure 16 shows the worn surface micrographs of glass fibre epoxy composite under different applied loads using different abrasive paper grades. Figure 16a shows that the worn surface of the glass epoxy composite is occurred in clear forms of the damages represented by plastic deformation, pull-out, detachment, and the breakage of the fibres which are due to the brittleness glass fibres. At grades of 80 and 400 (Fig. 16b, c), the ends of the glass fibre are exposed to the stainless steel counterface. This exposure subjects them to high shear loading which in turn causes

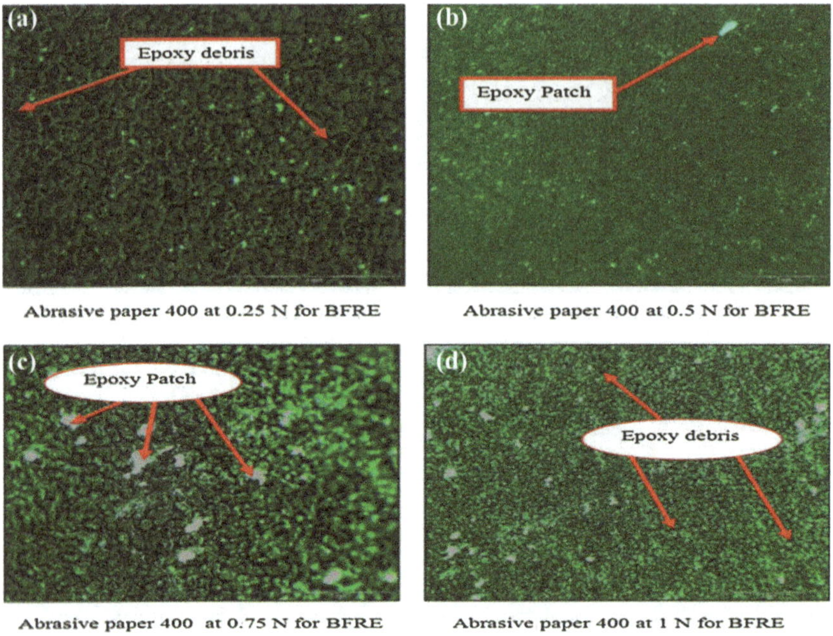

Fig. 16 Micrographs of the abrasive paper when the bamboo fibre reinforced epoxy tested under different applied loads using G400

the damage. At high applied loads, there is a removal of resinous areas followed by the pull-out of the fibres. This could be due to weakness in the interfacial bonding between the fibre and the polyester matrix. Similar results have been reported for polyester composites reinforced with glass fibre [3, 9]. In general, the micrograph of the worn surface of the glass epoxy composite shows different damage features for the three types of abrasive papers for all applied loads. The exposed fibres may contribute towards the increase in the material lost from the composite surface by the increasing coefficient of friction as mentioned in the previous frictional performance of glass, which in turn increases the wear rate. In regards to surface observation on the abrasive paper grades for the glass epoxy composite under microscope, similarly to other findings reported with bamboo epoxy composite and neat epoxy and presented in Fig. 17a–c.

A sample of micrographs of the worn surfaces for the neat epoxy and its composites under an applied load of 1 N using abrasive paper grade of 400 are presented in Fig. 18. For the neat epoxy, plastic deformation can be seen on the epoxy surface in the sliding process, i.e. a softening process for polyester resin occurred in response to the interface temperature. The effect of the interface

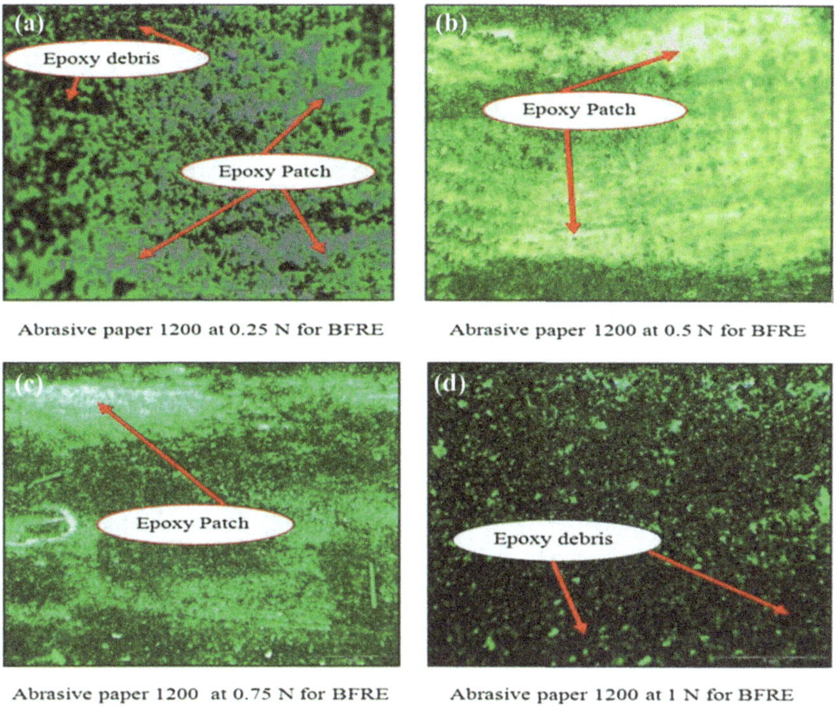

Fig. 17 Micrographs of the abrasive paper when the bamboo fibre reinforced epoxy tested under different applied loads using G1200

temperature on material removal is due to the friction loading causing more damage on the epoxy surface. In the bamboo epoxy composite, similar results can be observed. The softening process happens to the resinous areas which lead to the exposure of the bamboo fibre to the counterface causing the breakage of fibres over time (as mentioned in the optical microscopy of bamboo fibre section). However, the high damage occurring on the surface of glass epoxy composite can be seen under microscope. Of these damages, there are plastic deformation, material removal and pull-out of fibres when the composite is subjected to the applied load of 1 N.

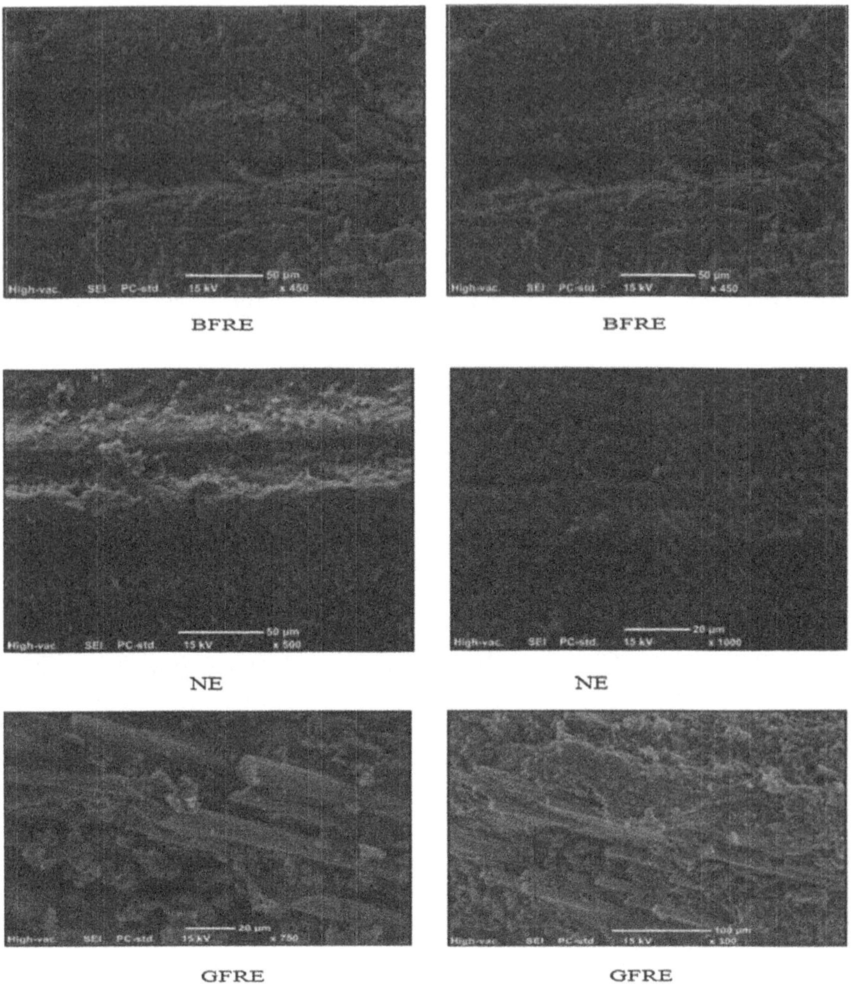

Fig. 18 Micrographs of the worn surface of the neat epoxy and its composites

4 Conclusions

In this study, a new epoxy composites based on the bamboo fibres were developed. The abrasive wear and frictional behaviour of the new composites were investigated with the neat epoxy and glass/epoxy composites. Different parameters and abrasive paper grades were used in the experiments. Based on the finding of the work, few points can be concluded as follows:

- The operating parameters (load and sliding distance) have great influence on the wear and frictional performance of all the composites. Steady state of the wear rate can be reached after few metres of the sliding distance with greater applied loads ≫ 0.5 N.

- SiC grades have equal influence on the wear and frictional behaviour of the composites with the operating parameters. At greater SiC grades, the wear rate reduces and the friction coefficient showed slight reduction as well. This was mainly due to the coverage of the high-grade SiC papers with the patches of the soft part (composites) which in turn generated a film on the SiC papers converting the abrasive into adhesive wear. This reduces the interaction between the asperities and then reduces the friction as well.
- Comparing the three developed composites (neat epoxy, bamboo/epoxy and glass/epoxy), bamboo fibres introduced better reinforcement to the epoxy in term of abrasive wear behaviour since it exhibited low wear rate compared to the glass and neat epoxy. It seems the softness and the strength of the bamboo fibres assisted to improve the surfaces during the rubbing and reduced the abrasiveness of the SiC papers. However, due to the fact that glass fibres are abrasive material, high wear rate achieved with the glass/epoxy since two high abrasive surfaces are in interaction which led to high material removal.

References

1. Chand N, Dwivedi U (2006) Effect of coupling agent on abrasive wear behaviour of chopped jute fibre-reinforced polypropylene composites. Wear 261(10):1057–1063
2. Chand N, Dwivedi U (2007) High stress abrasive wear study on bamboo. J Mater Process Tech 183(2–3):155–9
3. Chand N, Naik A, Neogi S (2000) Three-body abrasive wear of short glass fibre polyester composite. Wear 242(1):38–46
4. Chin C, Yousif B (2009) Potential of kenaf fibres as reinforcement for tribological applications. Wear 267(9):1550–1557
5. Chin C, Yousif F (2010) Influence of particle size, applied load, and fibre orientation on 3B-A wear and frictional behaviour of epoxy composite based on kenaf fibres. Proceedings of the Institution of Mechanical Engineers. Part J: J Eng Tribol 224(5):481–9
6. Edeerozey A, Akil HM, Azhar A, Ariffin M (2007) Chemical modification of kenaf fibers. Mater Lett 61(10):2023–2025
7. El-Tayeb N (2008) A study on the potential of sugarcane fibers/polyester composite for tribological applications. Wear 265(1):223–235
8. El-Tayeb N, Yousif B (2007) Evaluation of glass fiber reinforced polyester composite for multi-pass abrasive wear applications. Wear 262(9):1140–1151
9. Joshi SV, Drzal L, Mohanty A, Arora S (2004) Are natural fiber composites environmentally superior to glass fiber reinforced composites? Compos A Appl Sci Manuf 35(3):371–376
10. Nirmal U, Yousif B, Rilling D, Brevern P (2010) Effect of betelnut fibres treatment and contact conditions on adhesive wear and frictional performance of polyester composites. Wear 268(11):1354–1370
11. Okubo K, Fujii T, Yamamoto Y (2004) Development of bamboo-based polymer composites and their mechanical properties. Compos A Appl Sci Manuf 35(3):377–383
12. Ratna Prasad A, Mohana Rao K (2011) Mechanical properties of natural fibre reinforced polyester composites: jowar, sisal and bamboo. Mater Des 32(8):4658–4663

13. Sharma M, Rao IM, Bijwe J (2009) Influence of orientation of long fibers in carbon fiber-polyetherimide composites on mechanical and tribological properties. Wear. 267 (5-8):839–45
14. Singh Gill N, Yousif BF (2009) Wear and frictional performance of betelnut fibre-reinforced polyester composite. Proceedings of the Institution of Mechanical Engineers Part J: J Eng Tribol 223(2):183–94
15. Suresha B, Shiva Kumar K, Seetharamu S, Sampath Kumaran P (2010) Friction and dry sliding wear behavior of carbon and glass fabric reinforced vinyl ester composites. Tribol Int 43(3):602–9
16. Verheyde B, Rombouts M, Vanhulsel A, Havermans D, Meneve J, Wangenheim M (2009) Influence of surface treatment of elastomers on their frictional behaviour in sliding contact. Wear 266(3):468–475
17. Yousif B, El-Tayeb N (2008) Adhesive wear performance of T-OPRP and UT-OPRP composites. Tribol Lett 32(3):199–208
18. Yousif BF, El-Tayeb NSM (2010) Wear characteristics of thermoset composite under high stress three-body abrasive. Tribol Int 43(12):2365–71
19. Yousif BF, Nirmal U, Wong KJ (2010) Three-body abrasion on wear and frictional performance of treated betelnut fibre reinforced epoxy (T-BFRE) composite. Mater Des Eng 31(9):4514–21

Index

J.P. Davim (ed.), *Ecotribology*, Materials Forming, Machining and Tribology,
DOI 10.1007/978-3-319-24007-7

The manufacturer's authorised representative in the EU is Springer Nature Customer Service Centre GmbH, Europaplatz 3, 69115 Heidelberg, Germany. If you have any concerns regarding our products, please contact ProductSafety@springernature.com

Printed and bound by CPI Group (UK) Ltd, Croydon, CR0 4YY

22/07/2026

02174166-0014